Patterns and Systems of Elementary Mathematics

JONATHAN KNAUPP
Arizona State University

LEHI T. SMITH
Arizona State University

PAUL SHOECRAFT
Arizona State University

GARY D. WARKENTIN
Pacific College, Fresno

Houghton Mifflin Company   Boston

Atlanta   Dallas   Geneva, Ill.   Hopewell, N.J.   Palo Alto   London

1983

Patterns and Systems of Elementary Mathematics

Contents

Preface

Make the arrangement in Figure 1 with match sticks, toothpicks, or similar objects.

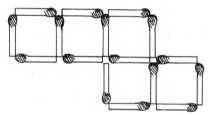

Figure 1

Can you change the positions of exactly two match sticks to make four squares instead of five? You must use the two match sticks somewhere without having two sticks on one side of a square. There are two ways of approaching this problem. The usual way is to try moving sticks randomly into various configurations. If this does not produce results, you will be forced to either abandon the puzzle or figure out a deductive solution.

One "smart" approach would be to reason that since there are a total of 16 match sticks, the four squares would not share any sides. Therefore, we would need to move sticks so there would be no adjacent squares. The solution to this puzzle is given at the bottom of page 16.

If you actually made the given arrangement with objects and attempted to move the sticks to solve this puzzle, you are already "in the groove" and should find this experience-oriented textbook enjoyable and worthwhile. If you are reading the preface but are not participating, there is still hope for you. You most likely believe that mathematics is just doing homework problems like the example the teacher did in class. Please join us and be actively involved. We want you to experience the satisfaction of participating in some "real" mathematics. You see, we believe that mathematics is a way of thinking, a way of doing, a way of organizing, and a way of predicting. Our experiences using the activities of this textbook indicate that if students have the right attitudes and the right interactions with their fellow students and instructors, they can do some clever and creative mathematical activities.

To the prospective teacher

You are probably the most important audience we have. After all, if you have not experienced the joy and pleasure of thinking mathematically, how can you provide school children with such experiences? How can you concentrate on the mathematical thinking of 30 children in the usual classroom if you have not had firsthand experience with such thinking yourself? We have tried to include only topics that are directly applicable to the elementary school mathematics program in grades K through 8.

There are many ways we could have exemplified the various topics in the text. We have chosen approaches that are similar to those of elementary school mathematics programs. Geoboards, balances, colored rods, numeration blocks, and attribute shapes are the basic manipulative models we have selected. Each of these items has been a useful aid in classrooms for many years. The authors have found them extremely effective for providing a concrete, manipulative basis for understanding the skills and concepts of the elementary mathematics to which this text is directed.

To the liberal arts major

For many of you, this will be your only college-level mathematics course. You are here because you needed a math elective or you are interested in learning more about mathematics. You will find the text an appropriate orientation to what mathematics is really all about. It will provide you with a basis for understanding your own children's school experience in the years to come.

Since you will generate mathematics from activities, you will develop an appreciation for the art of mathematics. It may be hard for you to understand this since the concrete level of activity seems immature. But let us assure you that reflecting on your actions and conclusions will give you a more accurate perception of elementary mathematics than any other type of activity. We have tried to emphasize the act of doing original (for you) work in elementary mathematics.

To the inservice teacher

To be a superior teacher, you need good methods and a good subject-matter background. This text is primarily concerned with mathematics content. But we present it with materials and in a style appropriate for elementary school instruction. We feel if you will learn mathematics in the way we suggest, you will automatically approach teaching with a similar point of view. That is, you will listen more closely to what children say about their mathematics and you will watch what they do with materials. You will consider providing manipulative experience for the basic mathematical concepts and skills. You will look at written work in mathematics as a record of what was actually done with materials or in activities.

Most important, you will find that basic skills and concepts will be worth remembering when you have discovered them yourself through the use of manipulative materials. Meaningful learning is what you can accomplish in your classroom if you have learned in that way yourself.

To the instructor

When all is said and done, you make the difference in any course taught from this text. Your style and approach to teaching may or may not be adaptable to the textual materials. It really comes down to this: Can you adapt to this style of instruction which is designed for nonmathematicians?

Our contention is that mathematicians think quantitatively at a level considerably more abstract than the nonmathematician. Although an abstract course in appreciation of mathematics has merit for non-mathematicians, our intent is to serve a different purpose. We believe our approach to mathematics is especially appropriate for prospective teachers and others who are interested in the conceptual development of basic mathematics.

For this textbook to really work, you need to make the students, with their discoveries and conjectures, the central issue of all classes. One-hour class sessions often consist of small groups of students explaining, discussing, and even arguing, with the instructor moving about the room responding to questions. This can be nerve racking for both the instructor and student, but you will soon become aware that the students truly are thinking. Your job is to suggest avenues for checking conjectures, to ask questions, and to explain why some of their "proofs" are valid or invalid.

You must also accept conjectures from everyone. This almost forces you to conduct many classes in a laboratory or small-group discussion mode. As you know, discussions in a class of 40 students are usually between the same few students and the teacher. We cannot emphasize enough the importance that *every* student be encouraged to make discoveries and conjectures. You can increase the students' opportunities for making conjectures by breaking the class into small groups.

Finally, the establishment of the proper attitude towards the text must be accomplished early in the course. The beginning chapters establish a mental set and general approach on which the later chapters are based. The chapters on whole numbers, integers, and rational numbers progress nicely if the previous chapters were completed successfully. With experience you will find a time schedule that fits your style of teaching and your objectives. Try the text in the way you feel is most appropriate, and then make adjustments in terms of the reactions of students.

We hope that you can enjoy using this approach as much as we have. We would appreciate suggestions and comments about your experiences with the text.

Several colleagues reviewed the manuscript for this text in its preliminary form and made many suggestions for its improvement that were ultimately incorporated; they are Jonathan Mudore, Arizona State University; Robert B. Dressel, Kent State University; John A. Dossey, State University of New York–Plattsburgh; and Stephen Krulik, Temple University. We would like to take this opportunity to thank the instructors and graduate students who made valuable contributions to this project—James Wiebe, Charles Riden, Lynnette Helmstedtler, James Hiebert, and William Rishel—and typists of the manuscript—Wanda Powell, Jane Cusick, Laurie Hughes, and Cyndie Osbahr.

Jonathan Knaupp
Lehi T. Smith
Paul Shoecraft
Gary D. Warkentin

1

Generating mathematical ideas

This text is quite different from the usual mathematics book. For your own psychological well-being, we advise you to read the Preface *before* you begin this chapter. The activities prescribed in this text are designed to generate ideas and conclusions, and it is essential that you understand the style and attitude you will need to successfully complete the assignments we shall generate. The Preface will help you get off to a good start.

1. Twister

Carefully remove Model *A* from the insert at the end of the book. Note that it is printed on heavy paper. Model *A* also appears on light paper in the appendix.

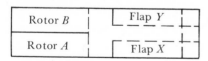

Figure 1

a. Fold flap *X* up.

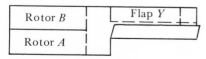

Figure 2

b. Fold flap *Y* down.

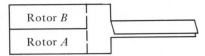

Figure 3

c. Fold flap *Z* over.

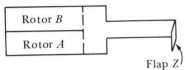

Figure 4

d. Fold rotor blade *A* to the right.

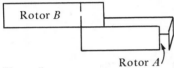

Figure 5

e. Turn the paper over.

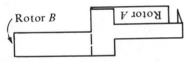

Figure 6

f. Fold rotor blade *B* to the right.

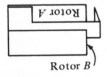

Figure 7

g. Hold at flap *Z* and "stroke" the rotor upwards.

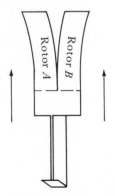

Figure 8

h. Hold out and drop.

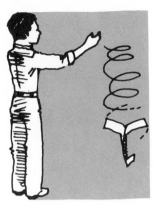

Figure 9

Now construct Model *A* from the light-weight paper model in the appendix. Hold the two twisters at the same height and release them at the same time. Which one falls faster? Which one rotates faster? Make them fall more slowly. Do they rotate faster now? Make them fall quickly, yet still rotate. Do they rotate faster now? Can you make them fall at the same speed?

Figure 10

1. Make three more twisters with various weights of paper. What can you conclude about the effect of the *construction material* on the flight of your twisters? Report your findings and conclusions in Table 1.

Rank the five twisters in order from Best (#1) to Worst (#5), and place check marks in the Rating column of Table 2.

2. Make an assortment of five twisters, ranging from small to large, of the *same type* of *materials*. Compare the effects of *size* on the flight of your twister. Report your findings and conclusions with appropriate sketches in Table 3.

Table 1   Weight-of-paper rating

Descriptions of weights of paper used	Rating of twisters
Paper *A* (lightest weight)	
Paper *B* (next lightest)	
Paper *C* (middle weight)	
Paper *D* (next heaviest)	
Paper *E* (heaviest)	

Table 2   Numerical rating display for weight

Best	1					
	2					
	3					
	4					
Worst	5					
		A	*B*	*C*	*D*	*E*

Lightest   Heaviest

Weight of paper

Table 3   Numerical rating display for size

Best	1					
	2					
	3					
	4					
Worst	5					
		A	*B*	*C*	*D*	*E*

Smallest   Largest

Size of twister

3. Change the proportions of the twisters. Compare the flights of short, stubby twisters with the flights of long, skinny twisters. What can you conclude about the effects of the proportions on the flight of your twisters? Draw graphs to express your data.

4. Do all your twisters rotate in the same direction? Can you change the direction of the rotation? Can you slow the rotation? Describe your findings.

5. Gather data on some things that affect the flight of your twister. In graphical form express the information given by the data. Be sure to label your axis and supply sufficient information so that your graph is understandable to other people.

6. Design your own twister so that it will stay aloft for a long time. Bring it to class for comparison with other students' designs.

The appendix contains a recording sheet for calculating times in case you choose to use a stop watch to time your flights.

2. Cardinal

Remove Model *B* on light paper from the appendix and Model *B* on heavy paper from the insert at the back of the book. Construct one of them by folding on the dotted lines, beginning at corner *A*.

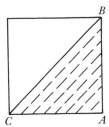

Figure 11

Continue folding until you come to the center line.

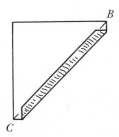

Figure 12

Now curve the folded part of the paper so that corner *C* meets corner *B* with the folds on the inside.

Figure 13

Corner *C* will fit into corner *B*. Fasten with some Scotch tape or glue. Place it on your hand as shown in Figure 14.

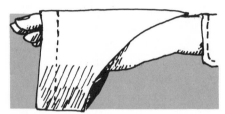

Figure 14

Now "flip" upwards as shown in Figure 15.

Figure 15

Now make the other Model *B*. Compare the flight of the two models. Make other models so that you can describe the effect of the following on the flight characteristics of the cardinal:

a. Type of paper
b. Size of model
c. Variations in design

1. Organize your data on these three variables and construct graphical displays of the information you gathered. Describe your conclusions verbally.

2. Design a cardinal that will fly a long distance. Bring it to class with data to substantiate your claims for your plane's flight.

Effective data gathering is a skill that must be learned from gathering data and making decisions about which variables are causing changes in outcomes. Getting accurate information on paper airplanes is difficult because it is almost impossible to control slight variations in design that cause major variations in flight characteristics.

Making sense out of complex data and being able to make accurate predictions about how variables are related is a major function of mathematics. This text should help you become more proficient at making accurate conjectures and predictions about patterns, relationships, and systems.

3. *Data gathering*

Gathering good data can be a difficult task, as you have just discovered in working with twisters and cardinals. The objective of the following exercises is to become aware of various aspects of data gathering and of that important concluding conjecture. Think about these questions as you attempt each of the following activities:

How much do you already know about the outcome of this activity?
How accurately do you need to measure to get "good" data?
How can organizing the data help in seeing relationships and patterns?
How can a formula be used to express these relationships and patterns?
How confident are you about your conjectures for the activity?

1. Use the second hand of a wrist watch to count your pulse in the following situations:

Situation	Pulse
a. Sitting quietly	☐
b. Right after running in place for 1 min	☐

Situation	*Pulse*
c. When you first wake up in the morning	☐
d. During a lecture class	☐
e. During this class	☐
f. When you are excited (you select the cause)	☐

What conclusions can you draw about how fast your heart beats?

2. Get a piece of string at least 10 ft long. Without measuring or putting it around yourself, quickly make a loop on the floor that is approximately the size of your waist.

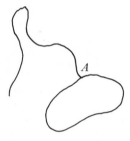

Figure 16

Cut at *A*. With the rest of the string, make a loop around your waist. This is your actual waist size. (Do not use a tape measure or other standard measuring device for this activity.) What is the difference between your actual waist size and your estimated waist size? Make a good estimate of this number and record how you arrived at it.

  Now have someone help you cut a string that is as long as you are tall. How many "waist" strings are in your "height" string? Again make an accurate estimate and record your procedures.

3. *Popcorn*   Prepare a popcorn popper for use in this activity. After the oil is heated, place $\frac{1}{4}$ cup of popcorn into the popper. After the corn is completely popped, measure out level $\frac{1}{4}$ cups of popped corn to determine how much bigger it got. Record your results for at least four trials in Table 4.

Table 4 Popped corn data

Trial	Unpopped $\frac{1}{4}$ cup	Popped $\frac{1}{4}$ cup
1	1	
2	1	
3	1	
4	1	
Total	4	

You may wish to invite several friends over for this activity. Have them bring various brands of liquid thirst quenchers. (Rootbeer?)

a. What can you conclude about the ratio of popped corn to unpopped corn?

b. Does the popcorn gain weight or lose weight?

c. What happens to the oil?

d. Design an experiment to determine the relative merits of two brands of popcorn.

4. *Famous number problem*   Take 10 toothpicks and cut them so that they are very close to 1 in. long. Hold them 1 ft above the parallel lines on page 371 of the appendix and drop them. Do this 10 times, and record in Table 5 the number of sticks that touch or cross a line. Divide this number into the total number of sticks dropped.

Table 5 Toothpick drop problem

Trial No.	1	2	3	4	5	6	7	8	9	10	Totals
Sticks dropped											
Sticks on line											

$$\frac{\text{Total sticks dropped}}{\text{Total sticks on a line}} =$$

Collect these same data from nine other members of the class. Add up the number of sticks dropped by all ten of you. Using Table 6, divide this figure by the total number of sticks on lines.

Table 6 Toothpick drop problem summary

Student No.	1	2	3	4	5	6	7	8	9	10	Totals
Sticks dropped											
Sticks on line											

If you did this carefully, your result should be an approximation of the number pi (π). How close was your ratio to 3.14?

5. *Tin-can pi* Find three tin cans of various sizes. Label them #1, #2, and #3. Do the following measurements and calculations to fill in Table 7.

Table 7 Tin-can data

Tin can	Diameter D	Height H	Circumference C	Side surface S	Top area T	$\frac{C}{D}$	$C \times H$	$\frac{C \times D}{4}$	$H \times T$
1									
2									
3									

a. Set the can on the graph paper found on page 373 of the appendix. Draw around the can.

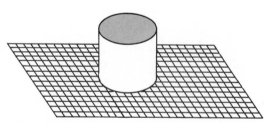

Figure 17

Count the squares across the outline at its widest point. Estimate to the nearest tenth of a square. Record under D in Table 7.

b. Now lay the tin can on its side so that it is lined up with the horizontal and vertical lines of the graph paper. Mark off the two ends of the can and count the squares between the ends to determine the height of the can. Record this information under H in Table 7.

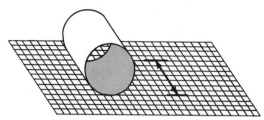

Figure 18

c. Lay the can on its side again. Mark a beginning point and roll one revolution. (Be sure to roll parallel to the lines of the paper.) Count the distance from starting point A to end point B. Record under C in Table 7.

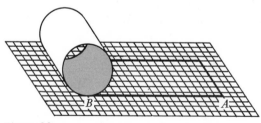

Figure 19

d. This time combine steps (b) and (c) to determine the surface area of the side of the can. Measure off height, then roll one revolution from A to B. Outline this rectangular region and count the squares. Estimate parts of squares. Record this number under S in Table 7.

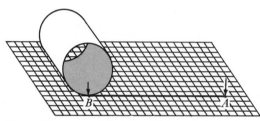

Figure 20

e. Set the can on end on the graph paper. Draw around the can. Remove the can and estimate the number of squares inside the region. Record this number under T in Table 7.

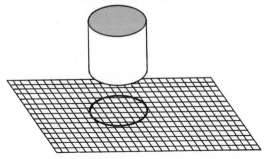

Figure 21

f. Now repeat this procedure for two other cans and record the results in Table 7.

g. For each can calculate the decimal approximation of C divided by D to two decimal places. Record this quotient under C/D in Table 7.

h. For each can calculate the decimal approximation of $C \times H$ to two decimal places. Record this product under $C \times H$ in Table 7.

i. For each can calculate the decimal approximation of $(C \times D)/4$ to two decimal places. Record this product under $(C \times D)/4$ in Table 7.

j. For each can calculate the decimal approximation of $H \times T$ to two decimal places. Record this product under $H \times T$ in Table 7.

k. Look at the numbers in the various columns of the table. If you counted squares carefully, there should be nearly equal numbers in certain pairs of columns.

S should be approximately the same as $C \times H$.

T should be approximately the same as $\dfrac{C \times D}{4}$.

$\dfrac{C}{D}$ should be approximately 3.14.

$H \times T$ is the volume.

l. See if you can quickly calculate the numbers missing in Table 8.

Table 8 Tin-can missing data

Tin can	Diameter D	Height H	Circumference C	Side surface S	Top area T	$\dfrac{C}{D}$	$C \times H$	$\dfrac{C \times D}{4}$	$H \times T$
4	4	2	12.5	25	12.5				
5	3	4							
6	2						18.8		

Formulas that describe constant relationships between variables are a major part of mathematics. Their wise and appropriate application not only saves problem solvers immense amounts of time but also provides more accurate calculating than actual measurement. Generating formulas from the activities for use in calculating areas, lengths, volumes, and relationships will be important in this text. Repeated experience of this nature will help you to understand formulas in general.

6. *Pendulum* This activity is purposely left unstructured to give you an opportunity to develop your own data-gathering procedures. There is only one task: Find out all you can about pendulums. Here are some sketches of various "hook-ups" to give you some ideas. Use sketches and tables to record what you did and to help you organize the data you gather.

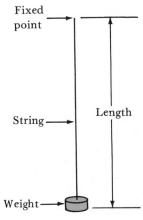

Figure 22

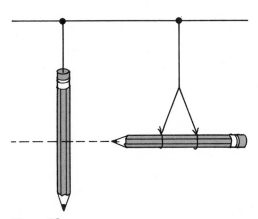

Figure 23

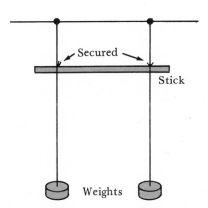

Figure 24

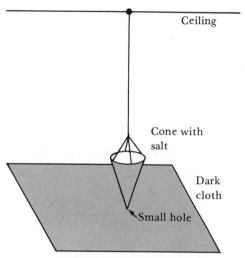

Ceiling

Cone with salt

Dark cloth

Small hole

Figure 25

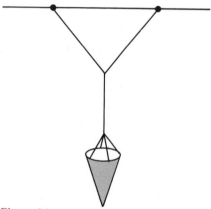

Figure 26

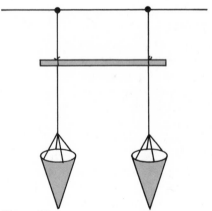

Figure 27

Summary

Each of the activities in this chapter deals with an important mathematical idea. Moreover, the activities embody a way of understanding that idea. The paper airplanes of Sections 1 and 2 demonstrate that a very complex task can be tackled with joy and pleasure by students with a wide range of experience and ability. The learner becomes involved with something he or she designs and builds. Personalizing the activity for the learner personalizes the mathematics derived from the activity. Even five-year-old children can construct, fly, and observe twisters and cardinals and become excited about the mathematics that describes their flight.

Counting your pulse is another example of a personalized mathematical experience. The uninteresting activity of simply counting pulse beats becomes quite fascinating when it is personalized. Whether an activity is interesting is related to the involvement of the learner.

Sometimes we are asked to find relationships when we have had no experience on which to base a good conjecture. The waist-estimate activity is an excellent example of how we can use mathematics to make a good estimate. If your waist is 30 in. around and it is in a circle-like shape, it will have a diameter of about $\frac{1}{3}$ its circumference: It will be about 10 in. across. Make the loop about 10 in. across and you have a good estimate of a 30-in. waist. Blindly guessing will usually produce a 50- to 60-in. loop. This string activity is also good for demonstrating means of finding remainders or ratios without actually measuring in standard units.

The popcorn activity illustrates that we can be very close to a situation but not look at it quantitatively. Most people do not realize that popcorn gets so much bigger. Even after they are told, they will want to actually pop some corn to find out for sure. This is a good chance to demonstrate the necessity of *firsthand* experience in accepting quantitative relationships. After a detailed analysis of the weight change in popcorn, a group of students might be able to write a complex formula to express their findings.

Perhaps the best activity in the chapter is designing an experiment to check which brand of popcorn is best. There are many variables, but it can be done with surprisingly consistent results. And eating the experiment afterwards seems to have positive effects on attitudes toward mathematics.

The famous number problem is an example of an activity that is magic! Someone very clever thought it up and it works (most of the time). We think this approach is appropriate when more natural means are not available. Although it is enjoyable to do the activity, the originator, not the learner, is the one who looks clever. We think pi is more appropriately generated by dividing the circumference by the diameter using measurements from tin cans. The chances are quite good that the learner will feel smarter than the originator (the tin can, in this case). This is a very important point: *It does the learner little good if someone else is always the one who looks smart!*

Studying pendulums is not difficult because the mathematics or physics is complex. It is rather the organization that is challenging. The results of this activity will indicate where you are in terms of dealing with the real world in a quantitative way.

Generally this chapter should set the stage for the attitudes we hope to generate within each learner. You are supposed to feel a degree of success in everything you attempt with this text. This is possible because answers alone are not the issue in the exercises that follow. It is the *process* you use to arrive at your "answers" that we want to improve. To gain the most from this text, you must actively participate by sharing your logical procedures (and the illogical ones) with your instructor and classmates. Only in this way can you improve your ability to understand the pattern and system of elementary mathematics.

Solution to match stick puzzle from the Preface

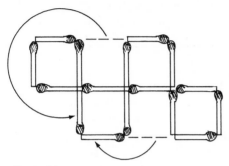

Figure 28

References

Junge, Charlotte. "Dots, Plots and Profiles." *The Arithmetic Teacher,* 16 (May 1969), 371–378.

Knaupp, Jonathan E., and Gary Knamiller. *Open Math Resource Book.* Arizona State University, Tempe, 1974.

Vervoort, Geradus. "Pie in the Street or How to Calculate π from the License Plates in the Parking Lot." *The Mathematics Teacher,* 58 (November 1975), 580–582.

2

Patterns and puzzles

When faced with solving a puzzle or finding a pattern, we all react in different ways. Too often our experience has taught us that the only objective is finding the correct answer. This chapter is more concerned with your *approaches* to problem solving.

As you do the exercises in this chapter, reflect on what strategy works best in finding the solutions. This strategy will become your tool for doing the exercises in the rest of the text. The basic questions are: How can you improve your strategies for solving problems? How can you increase the level of confidence you have in your predictions and conjectures?

1. Network problem

Figure 1 is called a *network*. This particular network consists of six *vertices*, nine *segments*, and four *interior regions*. The network in Figure 2 has seven *vertices*, eight *segments*, and two *interior regions*.

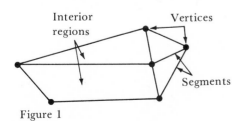

Figure 1

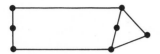

Figure 2

1. How many vertices, segments, and interior regions does this network have?

Figure 3

2. Find the number of vertices, segments, and interior regions for each of the networks in Figure 4 and record them in Table 1.

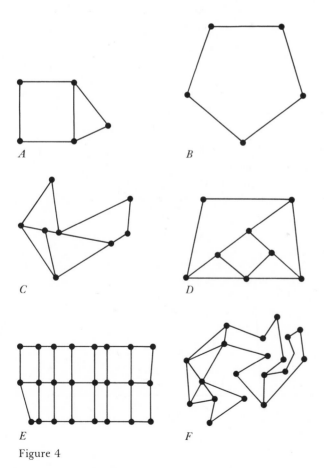

A

B

C

D

E

F

Figure 4

Table 1 Networks

	A	B	C	D	E	F
Vertices	5					
Segments	6					
Regions	2					

3. Make several networks and record the number of vertices, segments, and regions. Make a conjecture about the way in which the number of vertices, segments, and interior regions of networks are related. Try to find a counterexample which disproves your conjecture.

2. Circle problem

There are three dots on the circle in Figure 5. These dots are connected with line segments $\overline{AB}$, $\overline{BC}$, and $\overline{AC}$. The areas formed inside the circle are called *interior regions.* How many interior regions were formed?

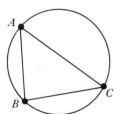

Figure 5

Using the same procedure, place dots on the circles in Figure 6 to complete Table 2.

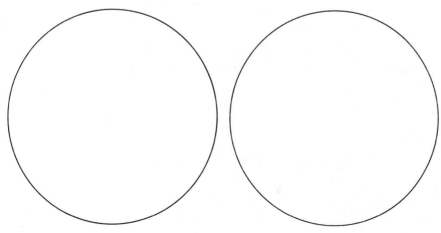

Figure 6

Table 2 Circle problem

Number of dots	1	2	3	4	5	6	7	8
Maximum possible interior regions								

If you place the dots in a symmetrical pattern, you will find that several line segments will pass through the same point. This reduces the maximum number of regions. Try several arrangements to be sure you make the maximum number of interior regions.

3. Number series

Determining patterns for number sequences is a very old form of mathematical recreation. The sophistication with which you make a prediction depends on your experience with algebraic notation. The following number series have patterns that can be described in words or with formulas. (Be sure to identify any symbols you use.)

1. Continue this pattern. Write out a general description of how it will continue.

$$1 = 1$$
$$1 + 3 = 4$$
$$1 + 3 + 5 = 9$$
$$1 + 3 + 5 + 7 =$$
$$1 + 3 + \cdots$$
$$1 + \cdots$$

2. Continue this pattern. Write out a general description of how it will continue.

$$1 = 1$$
$$3 + 5 = 8$$
$$7 + 9 + 11 = 27$$
$$13 + 15 + 17 + 19 =$$
$$21 + 23 + 25 + 27 + 29 =$$
$$31 + 33 + 35 + 37 + 39 + 41 =$$
$$43 + \cdots$$

3. Continue this pattern. Write out a general description of how it will continue.

$$1 = 1$$
$$1 + 1 = 2$$
$$1 + 1 + 2 = 4$$
$$1 + 1 + 2 + 3 = 7$$
$$1 + 1 + 2 + 3 + 5 = 12$$
$$1 + 1 + 2 + 3 + 5 + 8 = 20$$
$$1 + 1 + 2 + 3 + 5 + 8 + 13 = 33$$
$$1 + 1 + 2 + \cdots$$

4. Continue this pattern. Write out a general description of how it will continue.

$1 = 1$
$1 + 4 = 5$
$1 + 4 + 9 = 14$
$1 + 4 + 9 + 16 = 30$
$1 + 4 + 9 + 16 + 25 = 55$
$1 + 4 + \cdots$

4. Number triangle

The triangular array of numbers in Figure 7 is a famous number pattern known as *Pascal's Triangle*. There are many relationships between the various rows, columns, and diagonals of numbers. Continue this pattern for several more rows.

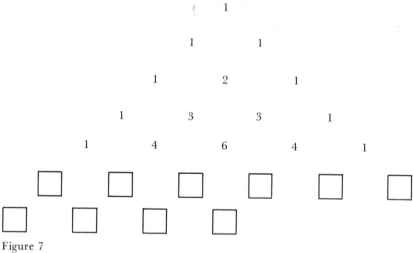

Figure 7

1. Describe two different ways in which this pattern generates itself.

2. What is the relationship between two adjacent numbers in one row and the number below them?

3. Explore other patterns in the triangle such as the sums of the numbers in each of the horizontal rows or the sequence of numbers in the diagonal columns, and make a conjecture about each of the patterns you discover.

5. *Number problems*

The following puzzles deal with some unusual *algorithms*, procedures by which a series of calculations result in a predictable outcome. The key to understanding these outcomes is in the structure of the numeration system itself. Try your skill at figuring out how these three procedures work.

1. To find the product of 47 and 43, we multiply 4 times 5 and place the answer to the left in the product space, then multiply 7 times 3 and place the answer to the right in the product space. Is this the correct product? Calculate the following examples in this way. Try to determine why it works and under what conditions it doesn't work.

EXAMPLES
$$
\begin{array}{r} 47 \\ \times\,43 \\ \hline 2021 \end{array}
\qquad
\begin{array}{r} 86 \\ \times\,84 \\ \hline 7224 \end{array}
$$

a. $\begin{array}{r} 64 \\ \times\,66 \\ \hline \end{array}$
 b. $\begin{array}{r} 72 \\ \times\,78 \\ \hline \end{array}$
 c. $\begin{array}{r} 35 \\ \times\,35 \\ \hline \end{array}$
 d. $\begin{array}{r} 53 \\ \times\,57 \\ \hline \end{array}$

2. Select any three different digits, such as 2, 5, and 7. Make all possible two-digit numbers using pairs of these three digits. You will always get exactly six different numbers. In this case you should have:

25, 52, 27, 72, 57, 75

Add these six numbers:

$25 + 52 + 27 + 72 + 57 + 75 = 308$

Find the sum of the original three digits:

$2 + 5 + 7 = 14$

Now divide the 308 by 14:

$\dfrac{308}{14} = 22$

a. Try the same procedure with these triples:

1, 3, 5

0, 4, 7

8, 9, 2

b. Make a conjecture about this procedure that applies to *any* three different digits.

c. Try to find a reason why this procedure works.

3. Choose any number such as 1984. Add the digits:

$1 + 9 + 8 + 4 = 22$

Continue adding the digits of the answer until the answer is a single digit:

$2 + 2 = 4$

Divide 1984 by 9:

$$\begin{array}{r} 220 \\ 9{\overline{)1984}} \\ 18 \\ \overline{18} \\ 18 \\ \overline{4} \end{array}$$

Note

Try this procedure with 17,760. Add the digits:

$1 + 7 + 7 + 6 + 0 = 21$

Add again:

$2 + 1 = 3$

Divide by 9:

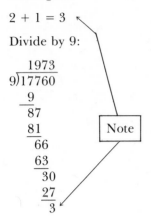

a. Select three other numbers and perform the same steps. Show your work.

b. Make a conjecture about this procedure.

c. Can you find exceptions to your conjecture?

6. *Magic squares*

The challenge of solving a seemingly complex problem in mathematics has different psychological implications for each of you. Many adults still have phobias about mathematics from school experiences. The following puzzles range in complexity from easy to very difficult. Try your luck (or skill) at solving these magic squares. Pay particular attention to a procedure for finding the solution rather than depending on blind luck.

1. In the 3×3 array, arrange the digits 1, 2, 3, 4, 5, 6, 7, 8, and 9 so that the sum of each column, each row, and each diagonal is 15. Write each number on a little square so that you can rearrange them conveniently. Describe the method you used to find a solution.

2. In the 4 × 4 array, arrange the numbers 1, 2, 3, 4, 5, 6, 7, 8, 9, 10, 11, 12, 13, 14, 15, and 16 so that the sum of each main diagonal is 34. Describe the method you used to find a solution. Find other arrangements that are clever.

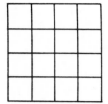

7. *People puzzles*

Your success in making sense out of mathematics is directly related to your ability to sort out relationships and patterns. An approach that works very effectively is to set up a record-keeping system so that you can identify what you know. The following three problems are good examples of how such a system can make a seemingly impossible puzzle quite manageable.

1. In the Valley State Bank, the positions of cashier, manager, and teller are held by Kaltenbach, Jones, and Mello, though not necessarily respectively. The teller, who is an only child, earns the least. Mello, who married Kaltenbach's sister, earns more than the manager. Determine which position each employee fills.

Table 3 Bank problem

	Cashier	Manager	Teller
Kaltenbach		X	
Jones			X
Mello	X		

2. Clark, Goldman, and Kondichook make their living as carpenter, painter, and plumber, though not necessarily respectively. The painter recently tried to get the carpenter to do some work for him, but was told that the carpenter was out doing some remodeling for the plumber. The plumber makes more money than the painter. Goldman makes more money than Clark. Kondichook never heard of Goldman. Determine the occupation of each.

Table 4 Workman problem

	Clark	Goldman	Kondichook
Carpenter	X		
Painter		X	
Plumber			X

3. The crew of a certain train consists of a brakeman, a conductor, an engineer, and a fireman. Their names are Art, John, Pete, and Tom. John is older than Art. The brakeman has no relatives on the crew. The engineer and the fireman are brothers. John is Pete's nephew. The fireman is not the conductor's uncle, and the conductor is not the engineer's uncle. What position does each man hold, and how are the men related?

Table 5   Train-workers problem

	Art	John	Pete	Tom
Brakeman	No Rel.			
Conductor		T's son		
Engineer			T's Brother	
Fireman				J's Dad

8. Famous train problems

Use logical reasoning rather than algebra to find the solutions to the following puzzles. Draw a sketch to illustrate each solution. The sketches should be to scale and carefully prepared. They will provide you with a visual representation for the situation. Reflect on how the sketches helped you solve the puzzles.

1. If a slow train can go up the hill at 1 mph and down the hill at 3 mph, what is its average rate for the entire trip?   *1½*

2. Two trains traveling in opposite directions meet on adjacent tracks. The first train is 1 mi long. The second train is 2 mi long. The first train is going 30 mph. The second train is going 15 mph. How long will it take for the trains to pass?   *4 min*

1 mi Train   ½ mile/minute
2 mi Train   ¼ mile/min.

3. If the trains in problem 2 were going in the same direction, how long would it take the shorter train to completely pass the longer train?

12 min

Summary

After doing all the puzzles in this chapter, you should appreciate the complexity of general problem solving. You just cannot become a good problem solver by depending on luck! You must have a logical, organized approach. And conclusions do not always need to be formalized. The pattern of the network problem can be stated: "The sum of the vertices and the regions is one more than the number of segments." That does the job nicely, although $V + R = S + 1$ is very neat and concise. The circle problem shows that we sometimes assume a pattern or relationship with very little data. Finding a concise description for this pattern is very difficult.

Number series demonstrate the considerable difference between being able to see patterns and being able to describe them in a neat, concise way. We think this is an important point to demonstrate to the learner since we sometimes portray that it is "easy to do abstract mathematics if you're smart like me." Spending a great deal of time generating the general formulas for the series would not be in keeping with the philosophy of this text.

Pascal's Triangle affords a good opportunity for creative pattern seekers. The applications of this neat arrangement could fill many pages, but our objective is to provide another opportunity for generating patterns and relationships.

Number problems can be used to point out the usefulness of understanding the numeration system. By representing the numbers symbolically and applying various laws of computation, seemingly "magic" procedures become quite understandable. But (and this is a most important point) whether you have proved it deductively does not affect its usefulness to you. If you consistently test the rule and it works all the time, you will feel just (almost?) as confident as the deductive mathematician.

We really had to include some magic squares in a chapter called "Patterns and Puzzles." Most find the 3×3 array very stimulating since it appears so simple. Students generally use a very crude "system" that depends on a lot of luck. But note the pleasure the solution gives regardless of how you find it. Don't overlook the difference between a strategy solution and a luck solution. They are both joyful, but the strategy solution has implications for the future.

People puzzles are an excellent example of how we make complex problems easy by introducing a system for recording information. In fact, isn't practical problem solving nothing more than taking what you know and placing it in an organized system to identify the solution?

Famous train problems illustrate the use of scaled sketches to recreate the famous situation. By drawing the sketches you can actually visualize what took place. From this you can find a logical procedure for generating a formal solution.

The variety of puzzles in this chapter should have helped you establish a mental framework for dealing with the rest of the text. In the next chapter you will begin dealing with more specific mathematical objectives, but you can use the same variety of approaches to develop *your* logical solution.

References

Hoffman, Nathan. "Pascal's Triangle." *The Arithmetic Teacher,* 21 (March 1974), 190–198.

Papy, Frederique. "Nebuchadnezzar, Seller of Newspapers: An Introduction to Some Applied Mathematics." *The Arithmetic Teacher,* 21 (April 1974), 278–285.

Sawada, Daiyo. "Magic Squares: Extensions into Mathematics." *The Arithmetic Teacher,* 21 (March 1974), 183–188.

Trigg, Charles. "Diagonally Magic Square Arrays." *The Arithmetic Teacher,* 20 (May 1973), 386–388.

Vaughn, Ruth K. "Investigation of Line Crossing in a Circle." *The Arithmetic Teacher,* 18 (March 1971), 157–160.

3

Geopatterns

The previous chapters should have made you aware that generating mathematics from real and abstract situations is a challenge. The real world does not present elementary mathematics in a simple, obvious manner. We would not want to subject you to discovering mathematical ideas or procedures only through such a complex approach.

We use structural models in the next eight chapters to provide visual and tactile experience for the basic ideas of the text. This chapter, for example, uses a geoboard as a source of patterns and sequences that generate and use procedures and concepts of elementary mathematics.

You can construct a geoboard or you can purchase one from any of the sources listed at the end of this chapter. Directions for constructing a 12 × 12 geoboard appear in the appendix.

We can only hope that you believe that this device will be a significant factor in later sections and chapters. Learn to use it as a calculator. Extensive experience with this approach convinced us to require all students to use the geoboard for solving the problems in the next two chapters.

1. Area

Finding areas of shapes on the geoboard is the essential skill for successfully completing activities in this chapter. With time and experience you will feel comfortable and confident. Be sure you know how to calculate the area of any shape with vertices at nails on the geoboard.

1. Using rubber bands, on your geoboard construct several closed shapes with different numbers of sides and record them in Figure 1. Classify and label them with the following names: triangle, quadrilateral, pentagon, hexagon, and octagon.

2. Construct the following triangles on your geoboard and record them in Figure 1: equilateral (all sides equal); isosceles (two sides equal); scalene (no sides equal); right (one 90° angle); acute (all acute angles); and obtuse (one obtuse angle). Is it possible to construct an equilateral triangle on the geoboard? Why?

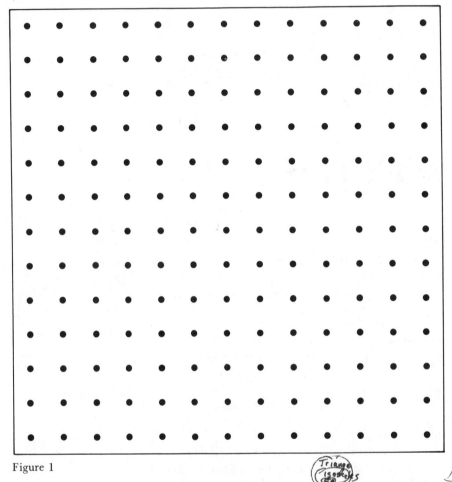

Figure 1

3. An equilateral triangle is also an isosceles triangle, and an isosceles triangle is a triangle. Determine a similar sequence of the following four-sided shapes: parallelograms, quadrilaterals, rectangles, and squares. We shall call the length between two nails on the geoboard, vertically and horizontally, *one unit*. (The diagonal length between two nails is *not* one unit.) The unit of area is a square with one unit on each side. The unit of area is referred to as *one square unit*.

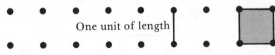

Figure 2

4. Compare the areas of shapes *A* and *B*; shapes *C* and *D*. Make a conjecture about how the diagonal divides the area of a square or rectangular shape.

$\div$ area in half

Based on your conjecture, determine the area of shape *E*.

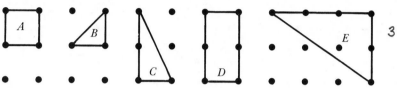

Figure 3

5. Study the two following techniques for finding the areas of more complex shapes.

   a. The area of a shape can be found by adding the areas of the parts into which the shape is subdivided.

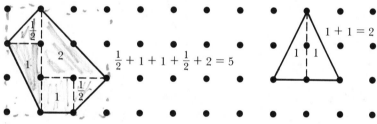

Figure 4

   b. It may be more convenient to find an area by subtracting the area of some small part from the known area of a larger part.

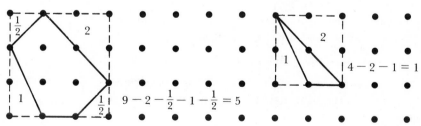

Figure 5

6. Construct the shapes in Figure 6 on your geoboard and determine their areas. Use additional rubber bands to help you. Write the number of units of the areas below.

   A __6__   *B* __10__   *C* __3½__   *D* __3__   *E* __3½__   *F* __9__   *G* __9__

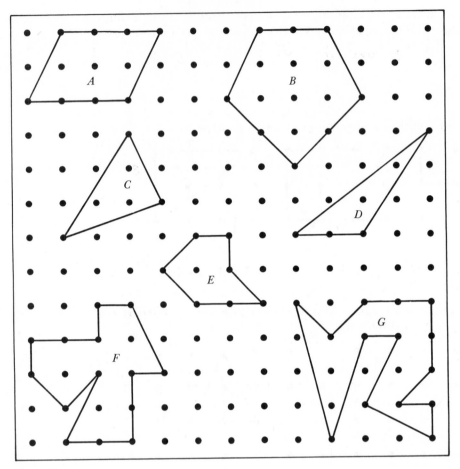

Figure 6

7. On the geoboard construct a five-sided shape with an area of 4 and record it in Figure 7.

8. Construct and record a square with an area of 2. (*Hint*: A square doesn't have to be vertical and horizontal on the geoboard.)

9. Construct and record a three-sided shape with the least possible area.

10. Construct several polygons that touch 8 nails and have 2 interior nails. Make a conjecture about the area of all figures touching 8 nails and having 2 interior nails. Record their areas.

Figure 7

11. Construct some other shapes, all touching the same number of nails and having the same number of nails in the interior. How do their areas compare?

Pic's Theorem

Border
Interior
Area

2. Squares

On your geoboard continue the sequence of squares shown in Figure 8. Note: The first square is part of the second square, and so on.

1. Find the area of the first 8 squares in the sequence.

a. __1__ b. __4__ c. __9__ d. __16__ e. __25__ f. __36__ g. __49__ h. __64__

2. How is the length of the sides of each square related to the area of the square?

$\sqrt{area}$

3. What is the increase in area between each of the squares in the sequence?

a. __3__ b. __5__ c. __7__ d. __9__ e. __11__ f. __13__ g. __15__ h. __17__

$1 = 1$

$1+3 = 4$

$1+3+5 = 9$

$1+3+5+7 = 16$

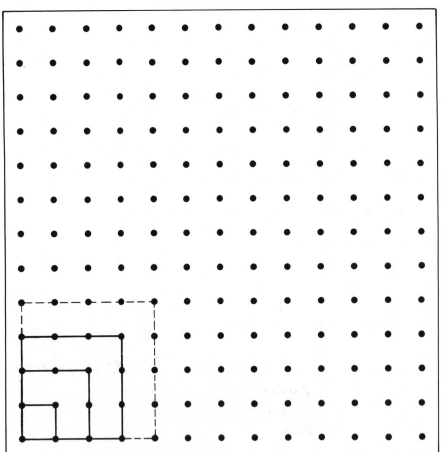

Figure 8

4. On your geoboard continue the sequence of shapes shown in Figure 9. Find the area of the first 8 squares in the sequence.

a. __2__ b. __8__ c. __18__ d. __32__ e. __50__ f. __72__ g. _____ h. _____

Make a conjecture about the sequence of areas. Compare this sequence with the one in exercise 3.

$$\left(\# \text{ segments per side}\right)^2 \times 2$$

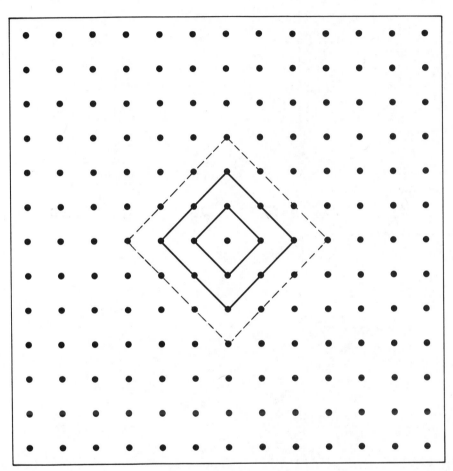

Figure 9

5. Continue the sequence of shapes in Figure 10 and find the area of the first 8 squares in the sequence.

a. ___2___ b. ___4___ c. ___8___ d. ___16___ e. ___32___ f. ___64___ g. ___128___ h. ___256___

Make a conjecture about the sequence of areas.

Each square is doubled.

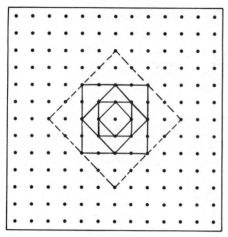

Figure 10

6. Continue the sequence of shapes in Figure 11 and find the area of the first 8 squares in the sequence.

a. ___2___ b. ___5___ c. ___10___ d. ___17___ e. ___26___ f. ___37___ g. ___50___ h. ___65___

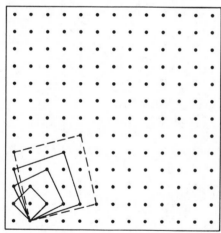

Figure 11

7. What is the increase in area between each of the squares in the sequence?

a. ___3___ b. ___5___ c. ___7___ d. ___9___ e. ___11___ f. ___13___ g. ___15___ h. ___17___

Make a conjecture about the sequence of areas.

$$a + 3 = b$$
$$b + 5 = c$$
$$c + 7 = d$$

3. Triangles

On your geoboard continue the sequence of shapes shown in Figure 12.

1. Find the area of the first 8 triangles in the sequence.

a. $\frac{1}{2}$ b. $\frac{4}{2}$ c. $\frac{9}{2}$ d. $\frac{16}{2}$ e. $\frac{25}{2}$ f. $\frac{36}{2}$ g. $\frac{49}{2}$ h. $\frac{64}{2}$

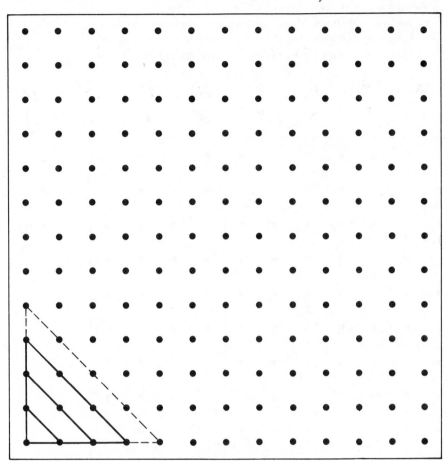

Figure 12

Make a conjecture about the sequence of areas. Is there a pattern?

$\frac{1}{2}$ (units per side)^2

side $\neq$ hypotenuse

2. Count the number of nails touching and enclosed in each of the triangles in Figure 12.

a. 3 b. 6 c. 10 d. 15 e. 21 f. 28 g. 36 h. 45

What pattern do you see in this sequence of numbers?

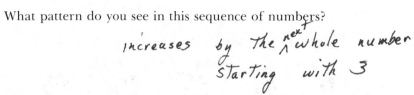

increases by the next whole number starting with 3

3. Continue the sequence of shapes shown in Figure 13. Find the area of the first 8 triangles and record them. Make a conjecture about the sequence of areas. Is there a pattern? Place the height below the corresponding area of each of the triangles. What relationship is there between the heights and the areas?

Area a. *2* b. *4* c. *6* d. *8* e. *10* f. *12* g. *14* h. *16*

Height a. *1* b. *2* c. *3* d. *4* e. *5* f. *6* g. *7* h. *8*

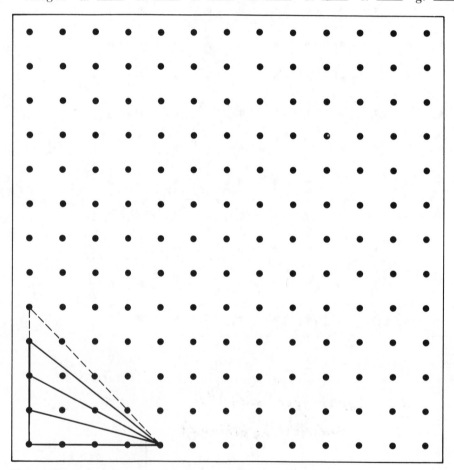

Figure 13

4. Construct another sequence of right triangles with a different base to check your conjectures.

5. Continue the sequences *A* and *B* in Figure 14. Find the area of the first 6 isosceles triangles in the sequences.

Sequence *A*   a. _1_   b. _2_   c. _3_   d. _4_   e. _5_   f. _6_

Sequence *B*   a. _2_   b. _4_   c. _6_   d. _8_   e. _10_   f. _12_

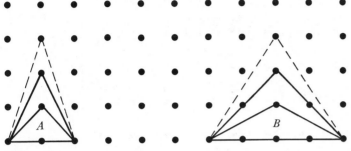

Figure 14

6. Construct and record a sequence of isosceles triangles with the following areas: 3, 6, 9, 12, 15, 18.

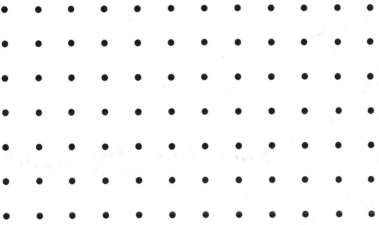

Figure 15

7. Make a conjecture about the relationship between the base, height, and area of *any* triangle. Try to find a triangle to disprove your conjecture.

$$A = \tfrac{1}{2} (b \times h)$$

8. Continue the sequence of shapes shown in Figure 16. This type of transformation is called a *shear*. Find the area of the first 7 triangles in the sequence. Make a conjecture about the shear transformation.

a. _10_   b. _10_   c. _10_   d. _10_   e. _10_   f. _10_   g. ____

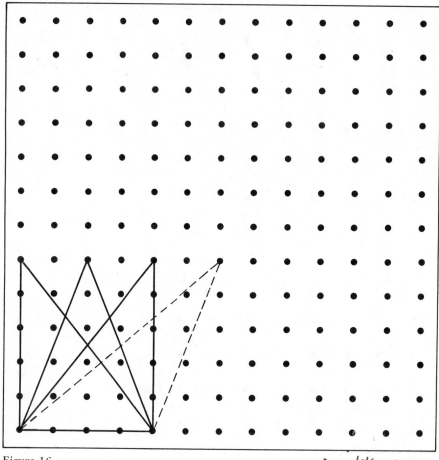

Figure 16

Base + height same
Area is constant

9. Construct another sequence of triangles with a different length of base. Do the shear transformation and record.

    a. _____   b. _____   c. _____   d. _____   e. _____   f. _____   g. _____

Does the conjecture you made for exercise 8 hold for the sequence of exercise 9?

10. Continue the sequences of triangles in Figures 17, 18, and 19. Find the area of the first 6 triangles in the sequences.

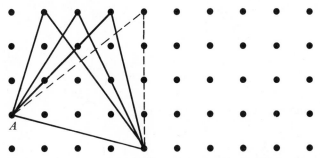

Figure 17

Sequence *A*   a. $\underline{13/2}$ b. $\underline{14/2}$ c. $\underline{15/2}$ d. $\underline{16/2}$ e. $\underline{17/2}$ f. $\underline{18/2}$

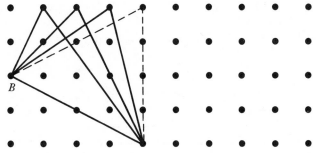

Figure 18

Sequence *B*   a. $\underline{10/2}$ b. $\underline{12/2}$ c. $\underline{14/2}$ d. $\underline{16/2}$ e. $\underline{18/2}$ f. $\underline{20/2}$

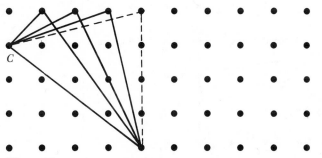

Figure 19

Sequence *C*   a. $\underline{7/2}$ b. $\underline{10/2}$ c. $\underline{13/2}$ d. $\underline{16/2}$ e. $\underline{19/2}$ f. $\underline{22/2}$

11. What is the pattern in each of sequences *A*, *B*, and *C*?

$A \qquad +\frac{1}{2}$

$B \qquad +\frac{2}{2}$

$C \qquad +\frac{3}{2}$

12. What is the effect of a shear when the base of a triangle is not horizontal or vertical?

13. Continue the sequence of rotating squares shown in Figure 20.

 a. Complete Table 1.

 b. Make a conjecture about the relationship between the area of each square and the distances to the upper corners of the original square.

 c. Relate your conjecture to the sides of the right triangles formed in the corners.

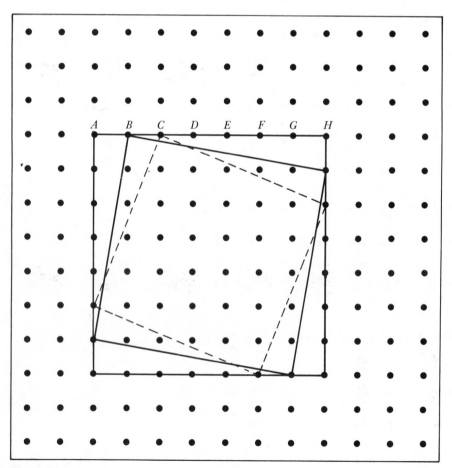

Figure 20

Table 1   Rotating square

	Units to left upper corner	Units to right upper corner	Area of square
A	0	7	49
B	1	6	37
C	2	5	29
D	3	4	25
E	4	3	25
F	5	2	29
G	6	1	37
H	7	0	49

14. The longest side of a right triangle is called its *hypotenuse,* and the other two sides are called *legs.* On your geoboard construct the sequence of right triangles from problem 13 and build squares on each of their sides as in Figure 21. Record their areas in Table 2.

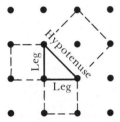

Figure 21

Table 2   Squares on triangles

	Area of square on one leg	Area of square on second leg	Area of square on hypotenuse
A	1	4	5
B	4	9	13
C	9	16	25
D	16	25	41
E	25	36	61

15. Make a conjecture about the relationship of the *areas* of the squares on a right triangle.

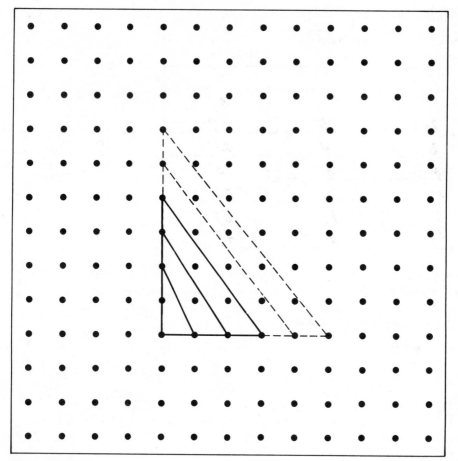

Figure 22

4. Line segments

The straight segments between two nails are called *line segments*. $\overline{AB}$, $\overline{CD}$, and $\overline{EF}$ are examples of line segments.

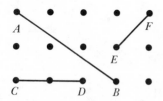

Figure 23

1. On your geoboard continue sequence *A* of Figure 24. Find the length of the first 7 line segments in the sequence.

a. _____ b. _____ c. _____ d. _____ e. _____ f. _____ g. _____

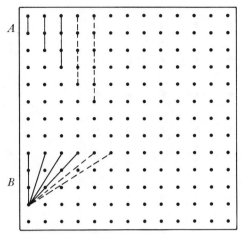

Figure 24

What is the pattern?

We have discovered that the sum of the areas of the squares on the two shorter sides of a right triangle is equal to the area of the square on the longest side. We also found that the length of the side of a square times itself equals its area. We shall call one of the two equal factors of a number the *square root* of the number. Consequently, the side of a square may be called the square root of its area. Because $2 \times 2 = 4$, for example, 2 is the square root of 4, which may also be written $\sqrt{4}$. There are not two equal rational numbers whose product equals 5 so that we express the square root of 5 as $\sqrt{5}$.

2. On your geoboard continue sequence B of Figure 24 and find the length of the first 7 line segments in the sequence.

a. $\sqrt{4}$ b. $\sqrt{5}$ c. $\sqrt{8}$ d. $\sqrt{13}$ e. $\sqrt{20}$ f. $\sqrt{29}$ g. $\sqrt{40}$

Make a conjecture about the sequence of lengths.

3. On your geoboard continue the sequence of line segments shown in Figure 25. Determine the length of the first 6 line segments in the sequence.

a. $\sqrt{5}$ b. $\sqrt{20}$ c. $\sqrt{45}$ d. $\sqrt{80}$ e. $\sqrt{125}$ f. $\sqrt{180}$

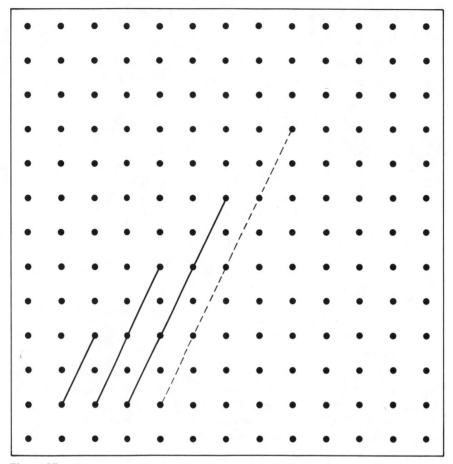

Figure 25

Because the second line segment is the side of a square whose area is 20, we can say its length is $\sqrt{20}$. However, since the second line segment is twice as long as the first line segment, it is simpler to say its length is twice $\sqrt{5}$ or $2\sqrt{5}$. Make a conjecture about the lengths in this sequence.

5. *Perimeter*

The distance around a shape is called its *perimeter*. In Figure 26 the perimeter of shape A is 8. Whole numbers can be added to irrational numbers, but such sums are expressed as a sum of the two types of numbers. The perimeter of shape B is written $7 + \sqrt{5}$. The perimeter of shape C is written $5 + 2\sqrt{2} + \sqrt{5}$.

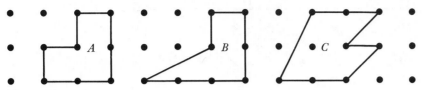

Figure 26

1. What is the perimeter of each of the shapes shown in Figure 27?

   D _____   E _____   F _____

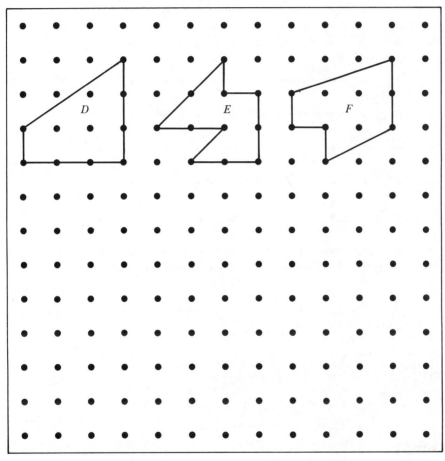

Figure 27

2. Find a shape with a perimeter of $4 + 2\sqrt{5}$ and record it in Figure 27.

3. Continue sequence *A* on your geoboard and record it in Figure 28. Find the perimeter of the first 6 triangles in the sequence.

   a. _____   b. _____   c. _____   d. _____   e. _____   f. _____

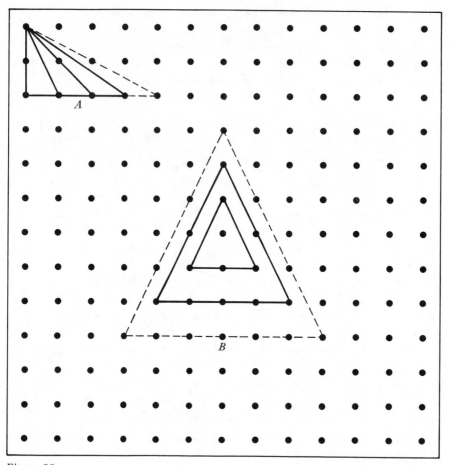

Figure 28

4. Continue sequence *B* on your geoboard and record it in Figure 28. Find the perimeter of the first 6 triangles in the sequence.

a. _____ b. _____ c. _____ d. _____ e. _____ f. _____

Make a conjecture about the sequence of perimeters.

6. Congruent shapes

1. Construct triangles *A* and *B* on your geoboard.

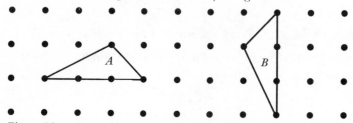

Figure 29

a. What is the sequence of lengths of the sides from shortest to longest for each triangle?

Triangle *A* $\sqrt{2}$ $\sqrt{5}$ 3

Triangle *B* $\sqrt{2}$ $\sqrt{5}$ 3

b. How are the sequences of sides related?

c. Find the area of triangle *A* _____ and triangle *B* _____ .

2. On your geoboard construct shape *C* and shape *D*.

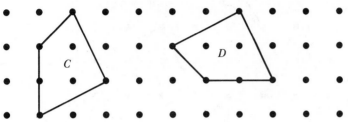

Figure 30

a. What is the sequence of lengths of adjacent sides for each shape?

Shape *C* 2 $\sqrt{2}$ $\sqrt{5}$ $\sqrt{5}$

Shape *D* 2 $\sqrt{2}$ $\sqrt{5}$ $\sqrt{5}$

b. How are the sequences of sides related?

c. Find the areas of shape *C* _____ and shape *D* _____ .

3. Shapes *A* and *B* are *congruent,* and shapes *C* and *D* are congruent. Write a definition for congruent shapes.

Same   size   same shape.

4. On your geoboard construct the shapes of Figure 31. Find 5 pairs of congruent shapes.

a. __A__ and __G__

b. __B__ and __J__

c. __C__ and __K__

d. __E__ and __F__

e. __E__ and __I__

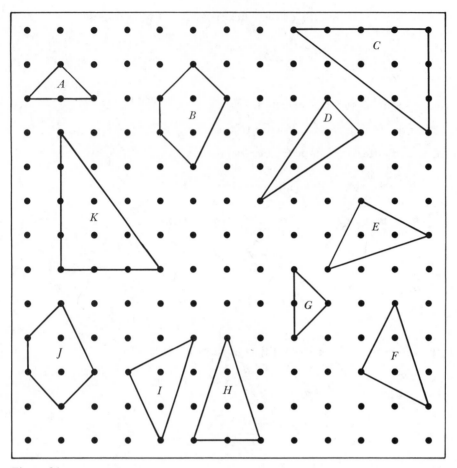

Figure 31

Fun

5. On your geoboard make the square of Figure 32. How many ways can you divide this shape into 2 congruent shapes? (Use only the nails on the geoboard as end points of segments.)

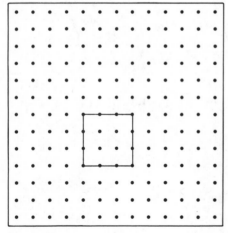

Figure 32

Fun

6. Divide shape E_1 into 2 congruent parts. Divide shape E_2 into 3 congruent parts. Divide shape E_3 into 4 congruent parts.

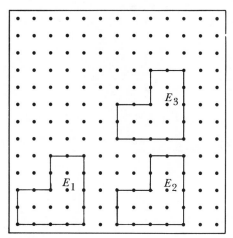

Figure 33

7. Similar shapes

1. On your geoboard construct triangles A and B.

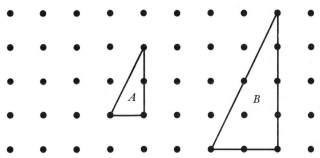

Figure 34

a. What is the sequence of lengths of the sides for each shape? (Begin with one side and move clockwise.)

Triangle A 1 2 $\sqrt{5}$

Triangle B 2 4 $2\sqrt{5}$

b. How are the sequences of sides related?

2. On your geoboard construct shapes C and D.

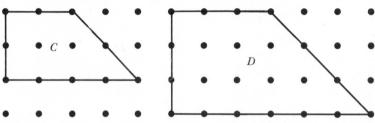

Figure 35

    a. What is the sequence of lengths of the sides for each shape?

       Shape C 2   2. $2\sqrt{2}$ 4

       Shape D 3   3 $3\sqrt{2}$ 6

    b. How are the sequences of sides related?

3. Shapes A and B are *similar,* and shapes C and D are similar. Write a definition for similar shapes.

4. Based on your definition, sketch a shape that is similar to shape E but not the same size.

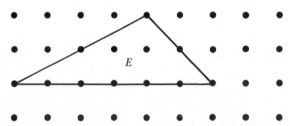

Figure 36

5. Explain why the rectangles on the sides of the right triangle in Figure 37 are similar.

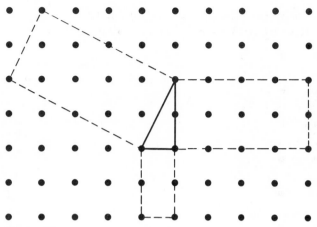

Figure 37

6. What is the area of each rectangle in Figure 37?

Small _____ Medium _____ Large _____

Can you see a relationship between these numbers?

7. Explain why the 3 triangles on the sides of the right triangle in Figure 38 are similar. Then find the area of each. Can you see a relationship between these 3 areas?

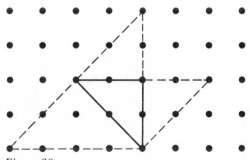

Figure 38

8. Find the areas of the L-shapes on the sides of the right triangle in Figure 39. Is there a relationship between these areas? Are the L-shapes similar?

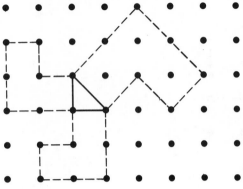

Figure 39

9. Finish constructing similar shapes on the right triangle in Figure 40. Does the sum of the areas of the shapes on the legs of the right triangle equal the area of the shape on the hypotenuse?

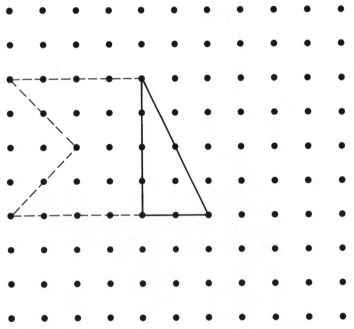

Figure 40

10. Finish constructing similar shapes on the sides of the right triangle in Figure 41. Explain why they are similar. Show that the sum of the areas of the two shapes on the legs is equal to the area of the shape on the hypotenuse.

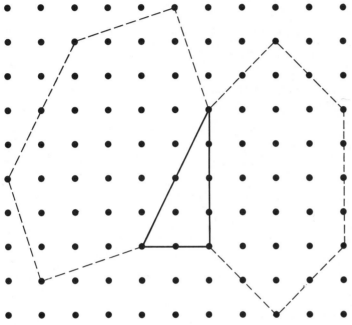

Figure 41

Summary

There are two basic outcomes that we were seeking in this chapter. The first is to demonstrate that you can deal with some rather complex mathematical ideas when they are based on orderly structural models such as geoboards. The second is to demonstrate that you do not need to deduce mathematical relations or properties to make them meaningful and useful. By now you should have made progress in thinking of mathematics as a logical, useful way of dealing with situations.

Our approach was to move from the complex real world to model situations that generate more clearly and efficiently the concepts and relationships that are the focus of this text. In fact, the next eight chapters use structural models almost exclusively. This permits us to let you discover the concepts and principles quickly and also gives you a convenient mental picture to associate with the content.

Although you may still feel uneasy with the approach we have chosen, we encourage you to continue manipulating the materials to model the situations we generate. Only through experience can you develop a lasting and successful style.

References

Aman, George. "Discovery on a Geoboard." *The Arithmetic Teacher,* 21 (April 1974), 267–272.

Edwards, Ronald. "Summing Arithmetic Series on the Geoboard." *The Mathematics Teacher,* 67 (May 1974), 471–473.

Hirsch, Christian. "Pick's Rule." *The Mathematics Teacher,* 67 (May 1974), 431–434.

Masalski, William. "An Open Ended Problem on the Geoboard." *The Mathematics Teacher,* 67 (March 1974), 264–268.

Mehl, William. "Where, on the Number Line, Is the Square Root of Two?" *The Arithmetic Teacher,* 17 (November 1970), 614–616.

Smith, Lewis B. "Pegboard Geometry." *The Arithmetic Teacher,* 12 (April 1965), 271–274.

Spaulding, Raymond. "Pythagorean Puzzles." *The Mathematics Teacher,* 67 (February 1974), 143–146.

Sullivan, John. "Polygons on a Lattice." *The Arithmetic Teacher,* 20 (December 1973), 673–675.

Geoboards can be obtained from the following sources.

Creative Publications
P.O. Box 10328
Palo Alto, CA 94303

Creative Teaching Associates
P.O. Box 293
Fresno, CA 93708

Houghton Mifflin Company
Customer Services Center
Wayside Road
Burlington, MA 01803

The Little Blue Whale Company
P.O. Box 27721
Tempe, AZ 85282

4

Finite systems

Most of the mathematics we use in our lives deals with numbers and the operations of addition, subtraction, multiplication, and division. In fact, we are so familiar with this mathematics that it is difficult to use it as the basis for developing an understanding of the nature of mathematics. A much better approach is to use finite systems that exemplify the basic concepts and properties of our number systems.

This chapter uses three concrete examples of finite mathematical systems to establish some fundamental concepts and properties of number systems and operations. Study the exercises carefully; this chapter is the foundation for many that follow.

1. Pattern system

In this section a geoboard makes a useful model for the mathematical system you will study. Place 16 rubber bands around 3×3 arrays as shown in Figure 6. Within these regions, make patterns according to the following rules:

Use exactly three (small) rubber bands for each array.

Figure 1

Connect a peg on the left to a peg on the right.

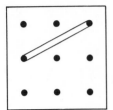

Figure 2

Do not use the center pegs.

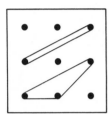

Figure 3

Use each peg exactly once.

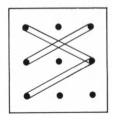

Figure 4

Do not connect any peg to two rubber bands.

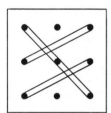

Figure 5

1. By using these rules, make as many different patterns as you can. Record them in Figure 6.

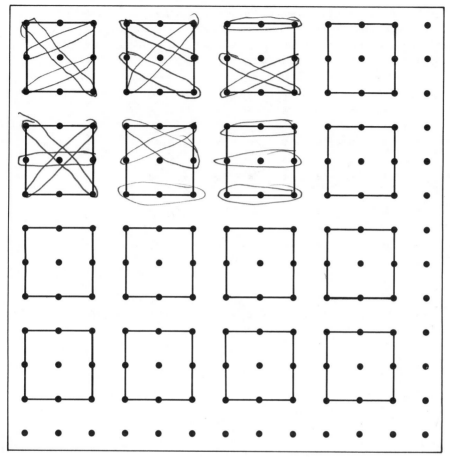

Figure 6

2. To communicate your findings, name the 6 patterns you found.

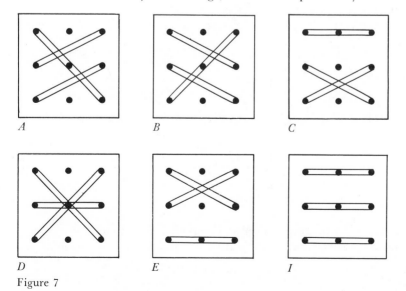

A   B   C

D   E   I

Figure 7

We shall consider all the configurations in Figure 8 the same pattern by ignoring their differences in size. In all cases the rubber bands must not be interlocked or hooked on an intermediate peg. The patterns must always be oriented horizontally. Vertical arrangements are not allowed for this system.

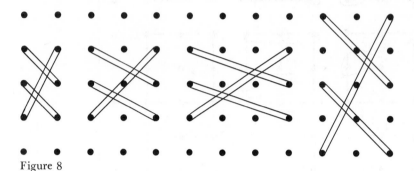

Figure 8

3. We can now use 3 × 3 arrays of pegs to make double patterns. Place one pattern on the left and center columns of pegs. Record this initial pattern in Figure 13.

Figure 9

Using these same rubber bands, extend them from the center pegs to form another pattern on the right. Record this second initial pattern in Figure 13.

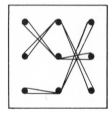

Figure 10

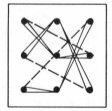

Figure 11

Release the rubber bands from the center column of pegs of each double pattern. Be sure to untangle rubber bands if they are crossed.

The resulting pattern is *B* given in Figure 12. Record the resulting pattern in Figure 13 with a different color from that used to record the initial patterns.

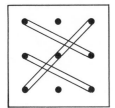

Figure 12

Do 15 more pairs of patterns and record them in the spaces provided in Figure 13.

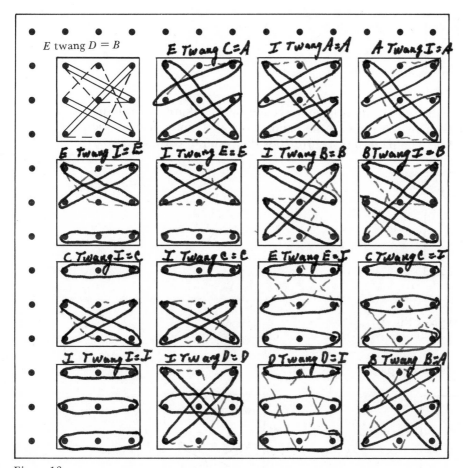

Figure 13

4. Using the 6 different patterns, we can generate a mathematical system. Place pattern *C* on the left part of the 3 × 3 array.

Figure 14

Now place pattern *A* on the right part of the 3 × 3 array by continuing the three rubber bands already on the board.

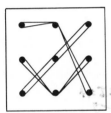

Figure 15

Now release the rubber bands from the center nails. Be sure there are no tangled bands. This releasing of the bands from the center nails is the *twang* operation. The pattern in Figure 16 is the result of *C* twang *A*.

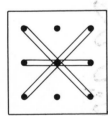

Figure 16

We would express this as

$C * A = D$

Here is another example:

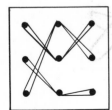

Figure 17

"Twang"
$B * E = C$

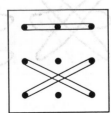

Return to Figure 13 and record the twang operation you performed on the sixteen 3 × 3 arrays.

$C * A = D$ $I * B = B$ $I * D = D$
$B * E = C$ $B * I = B$ $D * D = I$
$E * C = A$ $C * I = C$ $B * B = A$
$I * A = A$ $I * C = C$ $* \ =$
$A * I = A$ $E * E = I$ $* \ =$
$E * I = E$ $C * C = I$ $* \ =$
$I * E = E$ $I * I = I$ $* \ =$

5. In Table 1, record the results of all possible combinations of the 6 patterns in our system.

Table 1   Twang data

Right pattern
(Written second)

Left pattern
(Written first)

*	I	A	B	C	D	E
I	I	A	B	C	D	E
A	A	B	I	E	C	D
B	B	I	A	D	E	C
C	C	D	E	I	A	B
D	D	E	C	B	I	A
E	E	C	D	A	B	I

a. Select any 2 patterns from the set of 6. Put them on the 3 × 3 array and perform the twang operation.

Will you *always* end up with one of the 6 patterns?   yes

If you do, then we say that the set $\{A, B, C, D, E, I\}$ is closed under the twang operation. That is, under the twang operation, we cannot get any element outside the set.

b. Select any 2 patterns from the set of 6. Perform the twang operation on the 2 patterns. Now reverse the order of the 2 patterns and perform the twang operation.

Were the results the same? That is, did you end up with the same pattern in both cases?   yes

If you can change the order of any two elements of a set for an operation without changing the result, the operation is said to be _commutative_ on the set of elements. Is twang commutative on the set of 6 patterns?

c. Select any three elements of the set of 6 patterns. Write them in some order with the twang operation sign: $A * B * C$, for instance. Now perform the operation in two different groupings, doing the work inside the parentheses first:

$(A * B) * C = $ ___C___

$A * (B * C) = $ ___C___

Is the result the same?   *yes*

If the result will be the same for all triples of the 6 patterns, we say twang is *associative* on the set of 6 patterns. Does twang appear to be associative?   *yes*

d. Can you find a "do-nothing" member of the set of 6 patterns? That is, can you find a pattern that doesn't change other patterns when twanged with them?   *yes*

We call this do-nothing pattern the *identity* element for the twang operation. What is the identity for this system?   *I*

e. If you have an identity element for a system, you can look for special pairs of elements that produce the identity when twanged. Can you find 2 patterns in our 6 that can be twanged to produce the identity?

Each of these patterns is called the *inverse* of the other.

The inverse of pattern *A* is *B* because *A* twang *B* is *I*.

The inverse of *B* is: *A*
The inverse of *C* is: *C*
The inverse of *D* is: *D*
The inverse of *E* is: *E*
The inverse of *I* is: *I*

If each member of the set has an inverse, we say the system has the inverse property. Does our system have the inverse property?

2. Triangle system

Draw an equilateral triangle on a heavy piece of cardboard or construction paper. Cut the triangle out and label the corners on both sides of the triangle *A, B,* and *C,* as shown in Figure 18. Be sure to label each corner of the triangle on both sides with the same letter. See how many different positions you can generate by turning and flipping the triangle and setting it on the base line.

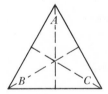

Figure 18

Here is a list of the six possible positions:

Figure 19

The first position can be achieved by starting with the triangle in Figure 18 and doing nothing to it. This will be called R_0 to signify no rotation. Another position can be generated by rotating $\frac{1}{3}$ of a rotation to position R_1; R_2 can be generated by $\frac{2}{3}$ of a rotation.

Figure 20

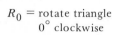

R_0 = rotate triangle
0° clockwise

Figure 21

R_1 = rotate triangle
120° clockwise

Figure 22

R_2 = rotate triangle
240° clockwise

F_A can be generated by flipping the triangle over the axis through corner A to the other side. F_B can be generated by flipping the triangle over the axis from corner B to the other side. F_C can be generated by flipping the triangle over the axis through corner C to the other side.

Figure 23

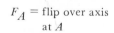

F_A = flip over axis
at A

Figure 24

F_B = flip over axis
at B

Figure 25

F_C = flip over axis
at C

Starting with the triangle in the standard position, we wish to perform first one element from the set and then another. We will indicate the process of "performing one operation and then another" by placing a "#" between the names of two of the elements from the set.

When flipping the triangle, always flip with respect to the specified corner no matter what position the triangle is in. Performing R_2 followed by F_A is the same as doing F_C alone, so we write:

$R_2 \# F_A = F_C$

1. Complete the following equations.

$R_1 \# F_A = \underline{F_B}$ $\qquad$ $F_B \# \underline{F_C} = R_2$

$F_A \# F_B = \underline{R_2}$ $\qquad$ $F_B \# \underline{F_B} = R_0$

$R_1 \# R_0 = \underline{R_1}$ $\qquad$ $R_0 \# \underline{F_C} = F_C$

$R_2 \# \underline{F_B} = F_A$ $\qquad$ $R_2 \# \underline{R_1} = R_0$

$R_2 \# \underline{R_2} = R_1$ $\qquad$ $\underline{R_1} \# R_2 = R_0$

2. There is one do-nothing motion. When it is combined with the other motions, it doesn't cause any changes. This is the R_0 motion. *(Identity)*

$R_0 \# F_A = \underline{F_a}$ $\qquad$ $F_B \# R_0 = \underline{F_b}$

$R_0 \# R_2 = \underline{R_2}$ $\qquad$ $R_1 \# R_0 = \underline{R_1}$

$F_C \# R_0 = \underline{F_c}$

Since R_0 doesn't change the other elements, we say R_0 is the <u>identity</u> element.

3. Fill out Table 2 to indicate the results of all possible combinations of the elements from our set.

Table 2   Triangle data

#	R_0	R_1	R_2	F_A	F_B	F_C
R_0	R_0	R_1	R_2	F_A	F_B	F_C
R_1	R_1	R_2	R_0	F_B	F_C	F_A
R_2	R_2	R_0	R_1	F_C	F_A	F_B
F_A	F_A	F_b	F_c	R_0	R_2	R_1
F_B	F_B	F_c	F_A	R_1	R_0	R_2
F_C	F_c	F_A	F_B	R_2	R_1	R_0

4. If two motions combined are the same as the (non)motion R_0, the two motions are *inverses* of each other.

EXAMPLE F_A is the inverse of F_A since $F_A \# F_A = R_0$.

  a. The inverse of F_C is: F_C

  b. The inverse of R_1 is: R_2

  c. The inverse of R_2 is: R_1

  d. The inverse of F_B is: F_B

  e. The inverse of R_0 is: R_0

Do any of the motions have more than one inverse? N_0

5. When combining two motions, does it matter what order they are performed in? For example, does $F_A \# R_1$ have the same result as $R_1 \# F_A$? *yes,* In general, if A and B are any two of the six motions, does $A \# B$ have the same result as $B \# A$? Is the $\#$ operation commutative? ~~yes~~) N_0

$$F_A \# F_B = R_2 \qquad F_b \# F_A = R_1$$

6. We may want to do several motions in succession, such as $(F_A \# R_1)$ $\# R_2$. Motions inside the parentheses should be combined first. Solve the following:

$(F_A \# R_1) \# R_2 = $ _F_A_ $\qquad R_2 \# (F_A \# R_0) = $ _F_C_

$(R_2 \# F_B) \# F_A = $ _R_0_ $\qquad R_0 \# (F_A \# F_B) = $ _R_2_

$(R_1 \# R_2) \# F_A = $ _F_A_ $\qquad R_2 \# (F_B \# F_A) = $ _R_0_

7. If every combination of elements from the set results in another element from the set, we say the operation on the set is *closed*. Is $\#$ closed on $\{R_0, R_1, R_2, F_A, F_B, F_C\}$?

$$yes$$

8. If $(A \# B) \# C = A \# (B \# C)$ for each three members of the set, we say the $\#$ operation is associative. Does $\#$ appear to be associative for $\{R_0, R_1, R_2, F_A, F_B, F_C\}$?

no

$$\left(F_A \# F_B \right) \# R_2 = R_2 \# R_2 = R_1$$

$$F_A \# (F \# R_2) = F_A \# F_A = R_0$$

3. Square system

Construct a square of cardboard and label it on both sides as shown in Figure 26. Be sure to label each corner on both sides with the same letters. The operation *move*, symbolized by @, is defined in Figure 27.

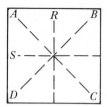

Figure 26

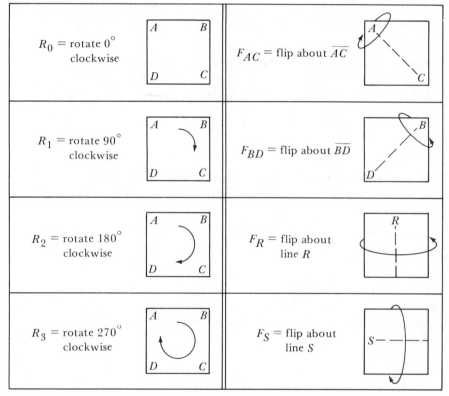

Figure 27

1. Fill in the corners of the squares in Figure 28 with the correct letters of the result.

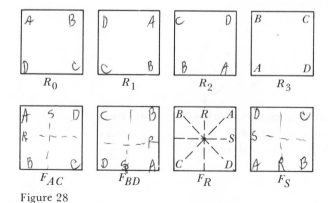

Figure 28

2. Using the same operation as in the triangle system, fill in Table 3. Remember to flip on the specified axis no matter what position the square is in.

Table 3 Square move data

@	R_0	R_1	R_2	R_3	F_{AC}	F_{BD}	F_R	F_S
R_0	R_0	R_1	R_2	R_3	F_{AC}	F_{BD}	F_R	F_S
R_1	R_1	R_2	R_3	R_0	F_R	F_S	F_{BD}	F_{AC}
R_2	R_2	R_3	R_0	R_1	F_{BD}	F_{AC}	F_S	F_R
R_3	R_3	R_0	R_1	R_2	F_S	F_R	F_{AC}	F_{BD}
F_{AC}	F_{AC}	F_R	F_{BD}	F_S	R_0	R_2	R_3	R_1
F_{BD}	F_{BD}	F_S	F_{AC}	F_R	R_2	R_0	R_1	R_3
F_R	F_R	F_{BD}	F_S	F_{AC}	R_1	R_3	R_0	R_2
F_S	F_S	F_{AC}	F_R	F_{BD}	R_3	R_1	R_2	R_0

a. Is @ closed on this set?

yes

b. Is @ associative on this set?

No

c. Does @ have an identity in this set?

yes R_0

d. Does each element have an @ inverse?

yes

e. Is @ commutative?

No

Summary

Finite systems have been used for many years to permit us to examine complete mathematical systems more easily. The number of elements that we must inspect is manageable, and it does not require complex proof procedures. Perhaps you can now use your experience from this chapter to establish the similar concepts and procedures of number systems. The following chapter is very similar to what you have just done, but the symbolism is more extensive and complex. Just as it was essential to use the manipulative model for the systems of this chapter, it is equally essential to use the model for the next chapter's exercises.

We consider Chapters 4 and 5 the foundation for developing an understanding of the systems of numbers and their corresponding numeration systems that are the basis of Chapters 6, 7, 8, and 9.

References

Meconi, L.J. "Discovering Structure through Patterns." *The Arithmetic Teacher,* 19 (November 1972), 531–533.

Oosse, William. "Properties of Operations: A Meaningful Study." *The Arithmetic Teacher,* 16 (April 1969), 271–275.

Usiskin, Coxford. *Geometry: A Transformation Approach.* Laidlaw Brothers, New York, 1975.

White, Paul. "An Application of Clock Arithmetic." *The Mathematics Teacher,* 66 (November 1973), 645–647.

5

Modular arithmetic

Understanding our number system is crucial to successfully working with numbers and arithmetic operations. Because our number system is an infinite set of numbers, it is often difficult to manage. Systems of modular arithmetic, however, deal with finite (and therefore more manageable) sets of elements that sometimes model our number system. Consequently, gaining an understanding of our number system is easier by examining these manageable systems.

In this chapter we generate systems of modular arithmetic with the colored strips included in the insert at the back of this book. In addition to being a finite number of elements, these strips allow us to observe how the elements and operations of our system function in concrete situations.

The following symbols will be used when referring to the colored strips:*

g = green	b = light blue	n = brown
k = black	o = orange	d = dark blue
r = rose	p = purple	t = turquoise
w = white	y = yellow	i = pink

1. End to end

Consider the set of strips and the following combining operation. Select any two strips and lay them end to end. Now find one strip that is just as long as these two laid end to end. For example, black laid end to end with light blue is the same length as purple.

---

* The sequence of colors in this book is the sequence used in the Ernest R. Duncan, Centimeter Rods for use with *School Mathematics: Concepts and Skills,* Houghton Mifflin Company, Boston.

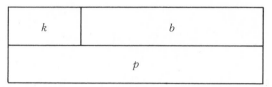

Figure 1

Orange laid end to end with brown is longer than the longest strip, pink, so we take only the small rose strip and call this the "result." Always place the longest strip first; then take what is left.

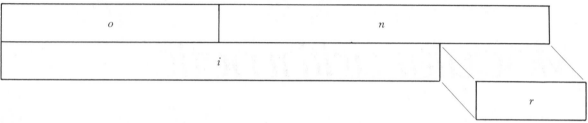

Figure 2

Here are some other examples. Check them out with your strips to be sure you understand the system.

p laid end to end with *r* is *d*
y laid end to end with *w* is *i*
i laid end to end with *k* is *k*
d laid end to end with *o* is *w*

Rather than write out "laid end to end" each time we talk about this operation, we shall use the ⊕ sign. Consequently, $o ⊕ y = k$ means orange laid end to end with yellow is black.

1. a. Find the results of the ⊕ operation on the set of strips and record them in Table 1.

   b. Is this system closed? Explain. **yes**

2. a. Find the results of the following:

   i. $(k ⊕ p) ⊕ w =$ ___ **g**

   $k ⊕ (p ⊕ w) =$ ___ **g**

   ii. $o ⊕ (b ⊕ r) =$ ___ **k**

   $(o ⊕ b) ⊕ r =$ ___ **k**

   iii. $(o ⊕ g) ⊕ y =$ ___ **r**

   $o ⊕ (g ⊕ y) =$ ___ **r**

   iv. $i ⊕ (d ⊕ r) =$ ___ **g**

   $(i ⊕ d) ⊕ r =$ ___ **g**

Table 1   End-to-end results

⊕	g	k	r	w	b	o	p	y	n	d	t	i
g	K	r	w	b	o	p	y	n	d	t	i	g
k	r	w	b	o	p	y	n	d	t	i	g	K
r	w	b	o	p	y	n	d	t	i	g	K	r
w	b	o	p	y	n	d	t	i	g	K	r	w
b	o	p	y	n	d	t	i	g	K	r	w	b
o	p	y	n	d	t	i	g	K	r	w	b	o
p	y	n	d	t	i	g	K	r	w	b	o	p
y	n	d	t	i	g	K	r	w	b	o	p	y
n	d	t	i	g	K	r	w	b	o	p	y	n
d	t	i	g	K	r	w	b	o	p	y	n	d
t	i	g	K	r	w	b	o	p	y	n	d	t
i	g	K	r	w	b	o	p	y	n	d	t	i

b. Recalling the patterns of the systems in Chapter 4, make a conjecture about the operation ⊕ on the set of 12 strips.   *associative*

c. Can you find a counterexample to disprove your conjecture?   *No*

3. a. Find the results of the following:

$n \oplus i =$ __*n*__ $y \oplus i =$ __*y*__

$i \oplus d =$ __*d*__ $i \oplus w =$ __*w*__

$i \oplus r =$ __*r*__ $b \oplus i =$ __*b*__

b. Make a conjecture about the pink strip and the operation ⊕.

identity

c. Is your conjecture *always* true?

yes

4. a. Fill in the following blanks:

$b \circledast \underline{\quad p \quad} = i$ $w \circledast \underline{\quad y \quad} = i$

$\underline{\quad o \quad} \circledast o = i$ $\underline{\quad t \quad} \circledast g = i$

$d \circledast \underline{\quad k \quad} = i$ $y \circledast \underline{\quad w \quad} = i$

b. Are b and p inverse elements in this system?

yes

c. Is there a strip that does not have an inverse? If so, what is it?

No

5. a. Find the results of the following:

    i. $r \circledast n = \underline{\quad i \quad}$ ii. $y \circledast g = \underline{\quad n \quad}$

      $n \circledast r = \underline{\quad i \quad}$ $g \circledast y = \underline{\quad n \quad}$

   iii. $p \circledast k = \underline{\quad n \quad}$ iv. $t \circledast w = \underline{\quad r \quad}$

      $k \circledast p = \underline{\quad n \quad}$ $w \circledast t = \underline{\quad r \quad}$

b. Make a conjecture about the operation $\circledast$ and the set of 12 strips.

commutative

c. Can you find a counterexample to disprove your conjecture?

no

6. A set of elements and an operation on the elements of the set are called a *group* if (a) the set is closed under the operation, (b) the operation is associative, (c) there is an identity element for the operation in the set, and (d) each element in the set has an inverse in the set.

    Do the set of 12 strips and the operation $\circledast$ form a group? *yes*

2. *Clock arithmetic*

Compare the system of the set of strips and the $\circledast$ operation to the system of hourly movements on a clock and their combinations. Because there are 12 possible hourly movements, there are 12 elements in this system. To combine 2 elements in this system, start at 12 o'clock and move the number of hours of the first element. From there move the number of hours of the second element. The result is the time at which you stop.

Figure 3

EXAMPLE   To combine 9 and 5, start at 12 o'clock and move 9 hr. From there more 5 hr. This motion results in the same motion as beginning at 12 o'clock and moving 2 hr. So the result is the two motions 9 and 5—which is the same as the motion 2.

1. Make a table of the results of the combination of hourly movements on the clock. This table should be like the one for the strips.   *1-12*

2. For example, look at the number 9 everywhere it appears in the clock face table. Then look at the symbol *n* everywhere it appears in the strip table. Are these two elements always in like positions? In general, how does the clock table compare with the table for the ⊕ operation on the strips?

1 - g	*5 - b*	*9 - n*
2 - k	*6 - o*	*10 - d*
3 - r	*7 - p*	*11 - t*
4 - w	*8 - y*	*12 - i*

3. Could we replace the table of the ⊕ operation on the strips with the clock system?

yes

4. If the operations and the elements of two systems are interchangeable, the systems are called *isomorphic*. We have seen that the strip system is isomorphic to, or is a model of, the clock system. In subsequent sections we shall see how the strips are also a model for other systems of arithmetic. In fact, one important focus of this book will be the isomorphisms we shall refer to through our use of models.

3. Stack 'em up and stretch 'em out

Consider another type of system using the set of strips and a different operation called *stack 'em up and stretch 'em out*. This operation is as follows: Select two strips and "stack 'em up" (Figure 4). Now make a floor with strips of the same color as the bottom strip. Be sure that the floor is as wide as the top strip (Figure 5). Discard the top strip and "stretch out" the floor strips. Now use the same procedure as with the laid-end-to-end operation, discarding multiples of the pink strip and taking the remaining strip as the result. The result of this operation on rose and blue is rose. We write $r \circledast b = r$ and say rose stack 'em up and stretch 'em out blue is rose (Figure 6). We designate this operation with the symbol $\circledast$.

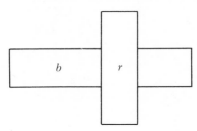

Figure 4

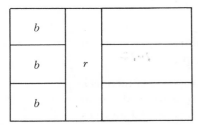

Figure 5

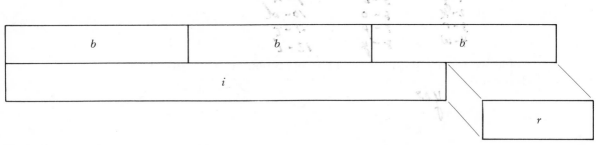

Figure 6

Check these examples with your strips.

$$k \circledast k = w \qquad b \circledast w = y$$
$$r \circledast w = i \qquad g \circledast n = n$$

1. Find the results of the following.

$$r \circledast i = \underline{\quad i \quad} \qquad b \circledast i = \underline{\quad i \quad}$$
$$w \circledast i = \underline{\quad i \quad} \qquad k \circledast i = \underline{\quad i \quad}$$

2. Make a conjecture about i and the operation $\circledast$. Can you think of a good reason why it is sometimes referred to as the "annihilator" for the $\circledast$ operation on the set of 12 strips?

$$i \circledast l = i$$

3. Record in Table 2 all possible results of the ⊛ operation on the set of strips.

Table 2   Discard twelves

⊛	g	k	r	w	b	o	p	y	n	d	t	i
g	g	K	r	w	b	o	P	y	n	d	t	i
k	k	w	o	y	p	i	K	w	o	y	d	i
r	r	o	n	i	R	o	n	i	r	d	n	i
w	w	y	i	w	y	i	w	y	i	w	y	i
b	b	D	R	y	g	o	t	w	n	K	p	i
o	o	i	o	i	o	i	o	i	o	i	o	i
p	p	K	n	w	t	o	g	y	R	d	b	i
y	y	w	i	y	w	i	y	w	l	y	w	i
n	n	o	r	i	n	o	p	i	n	o	r	i
d	d	y	o	w	K	i	d	y	g	w	K	i
t	t	d	n	y	p	o	b	w	r	K	g	i
i	i	i	i	i	i	i	i	i	i	i	i	i

4. a. Is the set $\{g, k, r, w, \ldots, i\}$ closed under the operation ⊛? *yes*

   b. Does it appear to be associative? *yes*

   c. Is there an identity element for ⊛ in the set? *yes (g)*

   d. Does every member of the set have a ⊛ inverse? *no — get a (g)*

      k, r, w, o, y, n, d, i

   e. Are the set of strips and the operation ⊛ a group? Why? *no — inverse*

5. Associate numbers with the lengths of each of the 12 strips. Identify an operation on these numbers that acts like ⊛ does on the set of strips. Make a table showing the results of performing the operation on the numbers. Show that this new system is isomorphic to the system of Table 2.

   X   operations + elements are interchangeable

6. More than one operation can be used on the set of strips at the same time. Using your set of strips and the operations ⊜ and ⊛, find the results of the following. Remember to do the operations inside the parentheses first.

a. $g \circledast (k \#r) =$ __b__ $(g \circledast k) \# (g \circledast r) =$ __b__

b. $r \circledast (p \#b) =$ __i__ $(r \circledast p) \# (r \circledast b) =$ __i__

c. $w \circledast (o \#g) =$ __w__ $(w \circledast o) \# (w \circledast g) =$ __w__

d. $b \circledast (y \#k) =$ __K__ $(b \circledast y) \# (b \circledast k) =$ __K__

7. Look carefully at the relationship between the four pairs of equations in exercise 6. A conjecture can be made about the equivalence of the left and right equations. If, in fact, this relationship holds for *all* triples of elements from our set, we say that $\circledast$ is distributive over $\#$. Is that a meaningful statement about this relationship?

8. Select the green, rose, white, light blue, and brown strips. Use the stack 'em up and stretch 'em out operation. However, discard multiples of the turquoise strip. Designate this operation by ©. Consequently, rose © light blue = white (Figure 7).

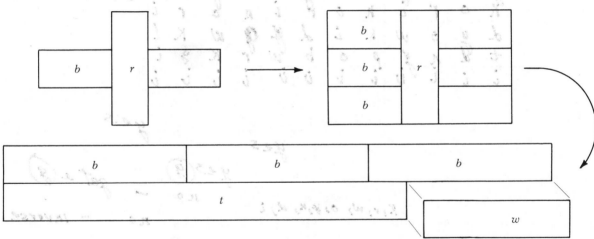

Figure 7

Find all possible results of the © operation on the green, rose, white, light blue, and brown strips, and record them in Table 3. Determine if these strips and the operation © form a group.

Table 3   Eleven

©	g	r	w	b	n
g	g	r	w	b	n
r	r	n	g	w	b
w	w	g	b	n	r
b	b	w	n	r	g
n	n	b	r	g	w

closed
commutative
w is the identity
each element has an inverse
strips & operation
form a
group

9. Select the green, black, rose, white, and light blue strips to use with the following two operations. The first operation is end to end discarding the light blue if the result is longer than light blue. Designate this operation by ⊕.

EXAMPLE $k \oplus w = g$

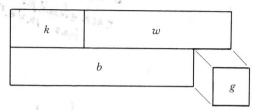

Figure 8

EXAMPLE $k \oplus r = b$

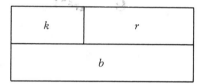

Figure 9

The second operation is stack 'em up and stretch 'em out, discarding multiples of the light blue strip. Designate this operation by ⊗.

EXAMPLE $r \otimes w = k$

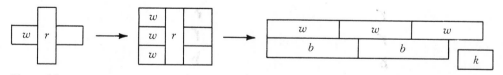

Figure 10

a. Check these results with your strips:

$$k \oplus r = b \qquad b \otimes r = b$$
$$r \oplus g = w \qquad k \otimes r = g$$
$$k \oplus w = g \qquad g \otimes w = w$$

b. Record the results of ⊕ and ⊗ in Tables 4 and 5.

Table 4   *Circle plus*

⊕	g	k	r	w	b
g	K	r	w	b	g
k	r	w	b	g	K
r	w	b	g	K	r
w	b	g	K	r	w
b	g	K	r	w	b

(handwritten) c. closed, Associative, b = identity, each element has inverse

Table 5   *Circle times*

⊗	g	k	r	w	b
g	g	K	r	w	b
k	K	w	g	r	b
r	r	g	w	K	b
w	w	r	K	g	b
b	b	b	b	b	b

(handwritten) d. × closed, × associative, g is the identity, e has ⊗ inverse except annihilator

c. Verify that this set of strips and the operation ⊕ is a group.

d. Exclude the annihilator and determine if this set and the operation ⊗ are a group. Explain your findings.

e. Find the results of the following:

$k \otimes (r \oplus g) =$ __r__ $(k \otimes r) \oplus (k \otimes g) =$ __r__

$w \otimes (r \oplus w) =$ __r__ $(w \otimes r) \oplus (w \otimes w) =$ __r__

$r \otimes (g \oplus w) =$ __b__ $(r \otimes g) \oplus (r \otimes w) =$ __b__

f. Make a conjecture about ⊕ and ⊗.

(handwritten) ⊗ distributes over ⊕

g. A set of elements and two different operations on the set are called a
field if (a) there is one operation that forms a group with the elements of
the set, (b) the second operation forms a group with the set without the
annihilator element, (c) each operation is commutative, and (d) the
second operation is distributive over the first.

4. Denoting inverses

(The exercises in this section refer to the operations $\oplus$ and $\otimes$ described in Section 3, problem 9.)

Tricky

We shall denote the $\oplus$ inverse of a strip with a raised minus sign *before* it.

EXAMPLE The $\oplus$ inverse of the black strip is denoted by ^-k and is equal to the rose strip: $^-k = r$.

We shall denote the $\otimes$ inverse of a strip with a raised -1 *after* it.

EXAMPLE The $\otimes$ inverse of rose is denoted by r^{-1} and is equal to black: $r^{-1} = k$.

b is identity

1. What are the following equal to?

   a. ^-g = **w**   b. ^-r = **K**

   c. ^-w = **g**   d. ^-b = **b**

   e. k^{-1} = **r**   f. w^{-1} = **w**

   g. b^{-1} = **doesn't exist** *ridiculous*   h. g^{-1} = **g**

2. Find the results of the following:

   a. $^-r \oplus ^-k$ = **K + r = b**

   b. $^-(r \oplus k)$ = **-(b) = b**

   c. $^-b \oplus ^-g$ = **b + w = w**

   d. $^-(b \oplus g)$ = **-(g) = w**

   e. $^-w \oplus ^-k$ = **g + r = w**

   f. $^-(w \oplus k)$ = **-(g) = w**

   g. Make a conjecture about adding $\oplus$ inverses.

   $$-a + -b = -(a + b)$$

3. Find the results of the following:

   a. $^-k \otimes ^-r$ = **g**

   b. $k \otimes r$ = **g**

   c. $^-w \otimes ^-g$ = **w**

   d. $w \otimes g$ = **w**

   e. $^-b \otimes ^-k$ = **b**

   f. $b \otimes k$ = **b**

   g. Make a conjecture about multiplying $\oplus$ inverses.

   $$-a \times -b = a \times b$$

4. Find the results of the following:

a. $^-r \otimes w = $ __r__

b. $^-(r \otimes w) = $ __r__

c. $k \otimes {}^-g = $ __r__

d. $^-(k \otimes g) = $ __r__

e. $^-r \otimes k = $ __w__

f. $^-(r \otimes k) = $ __w__

g. Make a conjecture about multiplying by a $\oplus$ inverse.

$$-a \times b = -(a \times b)$$

5. Find these $\otimes$ products:

a. $k \otimes r^{-1} = $ __$K \times K = w$__

b. $k^{-1} \otimes r = $ __$r \times r = w$__

c. $r \otimes g^{-1} = $ __$r \times g = r$__

d. $r^{-1} \otimes g = $ __$K \times g = K$__

e. $w \otimes r^{-1} = $ __$w \times K = r$__

f. $w^{-1} \otimes r = $ __$w \times r = K$__

g. $b \otimes k^{-1} = $ __$b \times r = b$__

h. $b^{-1} \otimes k = $ __cannot do, no $\otimes$ inverse for b__

i. Make a conjecture about the relationship between the pairs of products you just calculated. Pay particular attention to the last expression, $b^{-1} \otimes k = $ __$(a^{-1} \times b) = (a \times b^{-1})^{-1}$__ any elements

5. Factors and powers

(The exercises in this section refer to the operations described in Section 3, problem 9.)

When two or more strips are combined by $\otimes$, each of these strips is called a *factor* of the result.

EXAMPLE   White and rose are factors of black.

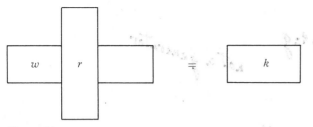

Figure 11

Find some sets of factors of the following strips:

a. Green   rose + black

b. Black   white + rose

c. Rose   white + green

d. White   rose + white

A result that has a set of factors that are all the same color is called a *power* of that color. We indicate the number of factors with a raised numeral (called the *exponent*) after the factor.

EXAMPLE   Black is a power of rose because $r \otimes r \otimes r = k$ (Figure 12), and we write $r^3 = k$.

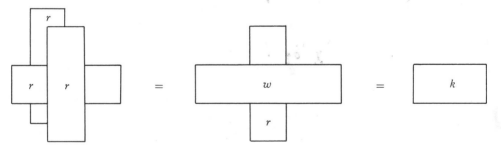

Figure 12

1. What are the powers of the rose strip?

a. $r^1 =$ __r__   b. $r^2 =$ __w__

c. $r^3 =$ __K__   d. $r^4 =$ __g__

e. $r^5 =$ __r__   f. $r^6 =$ __w__

2. Make a conjecture about the sequence of powers of rose. Is there a similar pattern in the powers of the black strip? If so, what is it?

r, w, K, g, r, w, K, g...

black
K, w, r, g, K, w, r, g

3. The powers of rose and black generate all the other strips in the system. Check whether white is a *generator*.

$w^1 = w$ $w^4 = g$

$w^2 = g$ *not a generator*

$w^3 = w$

4. Find the results of the following.

a. $r^2 \otimes r^1 =$ ___**K**___   b. $k^3 \otimes k^2 =$ ___**K**___

$r^3 =$ ___**K**___ $k^5 =$ ___**K**___

c. $w^1 \otimes w^3 =$ ___**g**___   d. $r^2 \otimes r^4 =$ ___**w**___

$w^4 =$ ___**g**___ $r^6 =$ ___**w**___

5. Make a conjecture about multiplying powers.

add exponents if bases are the same.

6. Use your conjecture to find the result of each of the following as a power of the strip being combined:

a. $w^3 \otimes w^0 =$ ___**w³**___   b. $k^0 \otimes k^4 =$ ___**k⁴**___

c. $r^0 \otimes r^2 =$ ___**r²**___   d. $r^1 \otimes r^0 =$ ___**r¹**___

7. Make a conjecture about the zero power and how it behaves with the $\otimes$ operation.

× by 1

6. Subtraction

Consider the new operation $\ominus$. To combine two strips with $\ominus$, place the first strip above the second. Then find a strip that will make the bottom strip as long as the top strip and call it the result.

EXAMPLE $w \ominus k = k$

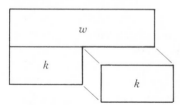

Figure 13

If the second strip is as long as or longer than the first, add a light blue strip to the first and find the strip that is as long as the first strip and the light blue strip.

EXAMPLE $k \ominus r = w$

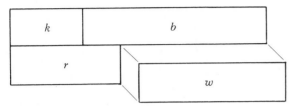

Figure 14

1. Find the results of the following.

   a. $r \ominus b$ = ___*r*___   b. $r \oplus {}^-b$ = ___*r*___

   c. $b \ominus w$ = ___*g*___   d. $b \oplus {}^-w$ = ___*g*___

   e. $g \ominus w$ = ___*k*___   f. $g \oplus {}^-w$ = ___*k*___

   g. $r \ominus g$ = ___*k*___   h. $r \oplus {}^-g$ = ___*k*___

   i. We call the operations $\oplus$ and $\ominus$ *inverse* operations. Examine each pair of the preceding examples, and make a conjecture about why these operations are called *inverse* operations.

   $$a - {}^+b = a + {}^-b$$

2. Find the results of the following:

   a. $r \ominus {}^-k$ = ___*b*___   b. $r \oplus k$ = ___*b*___

   c. $g \ominus {}^-w$ = ___*b*___   d. $g \oplus w$ = ___*b*___

   e. $b \ominus {}^-r$ = ___*r*___   f. $b \oplus r$ = ___*r*___

   g. Make a conjecture about subtracting a $\oplus$ inverse.

   $$a - {}^-b = a + b$$

7. Division

Consider the operation $\oslash$. To combine two strips with $\oslash$, place the first strip above the second and find how many strips of the second color are the same length as the first. Lay these second strips side by side and find a strip that is as long as they are wide. Call this strip the *quotient.*

EXAMPLE $w \ominus k = k$

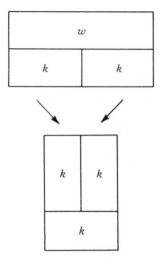

Figure 15

If you can't find the exact number of strips, add light blue strips to the first strip until you can.

EXAMPLE $k \ominus r = w$

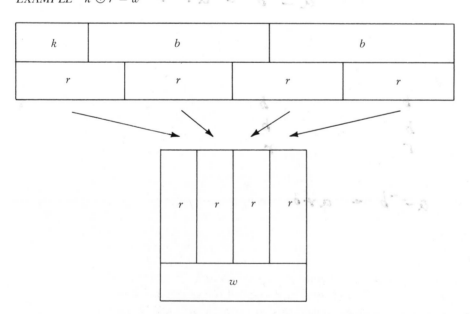

Figure 16

1. Find the results of the following.

    a. $r \otimes g^{-1} =$ ___*r*___   b. $r \oplus g =$ ___*r*___

    c. $w \otimes r^{-1} =$ ___*r*___   d. $w \oplus r =$ ___*r*___

    e. $b \otimes k^{-1} =$ ___*b*___   f. $b \oplus k =$ ___*b*___

    g. $r \otimes r^{-1} =$ ___*g*___   h. $r \oplus r =$ ___*g*___

    i. The operations $\otimes$ and $\oplus$ are inverse operations. Make a conjecture about why they are called *inverse* operations.

    $$a \times b^{-1} = a \div b$$

2. Find the results of the following.

    a. $(r \otimes w) \otimes r^{-1} =$ **k · k = w** b. $r \otimes (w \otimes r^{-1}) =$ **r · r = w**

    c. $(r \oplus w^{-1}) \oplus r =$ **k ÷ r = w** d. $r \oplus (w^{-1} \oplus r) =$ **r ÷ r = g**

    e. $(w \otimes r^{-1}) \oplus k =$ **r ÷ k = w** f. $w \otimes (r^{-1} \oplus k) =$ **w × g = w**

Summary

The chapter you have just completed deals with some of the most complex and demanding ideas in the entire text. It may be totally different from anything you have ever seen in mathematics. And although the notation looks familiar, it must be used very carefully and thoughtfully.

We would be surprised if you were able to do this chapter without depending on the colored strips as an aid. We have found that students who did not use the materials in previous chapters often have difficulty sorting out the meanings of the symbols, procedures, and concepts of this chapter.

Of course, we would not want you to remember much of how to do specific problems in this chapter. The idea is to firmly establish the meaning of numbers and operations, their properties, and the symbolic notation used to convey information. Too often we assume that number systems are simple and easy to understand and use. Actually they are quite complex and need to be carefully described and studied with manipulative models to be used with understanding. This is especially true for young children.

As you do the exercises in the next chapter, reflect on the development constructed here. The concepts are the same, but now you will be dealing with an infinite system of numbers and a new way to denote them.

References

Goldenberg, E. Paul. "Scrutinizing Number Charts." *The Arithmetic Teacher,* 18 (December 1970), 645–653.

Lyda, W. J., and Margaret D. Taylor. "Facilitating and Understanding of the Decimal Numeration System through Modular Arithmetic." *The Arithmetic Teacher,* 11 (February 1964), 101–103.

Mauro, Carl. "Developing an Understanding of Inverse Operations." *The Arithmetic Teacher,* 13 (November 1966), 556–563.

Plants, Robert. "Casting Out of Nines with Modular or Clock Arithmetic." *The Arithmetic Teacher,* 12 (October 1965), 460–461.

6

Base-five arithmetic

A major problem in teaching an understanding of elementary mathematics to adults is their familiarity with some terms and procedures. It is difficult to "forget" what we already know simply for the sake of instructional convenience. This is our best reason for dealing extensively here with base-five, whole number arithmetic.

The steps in this chapter are analogous to the development of the system of whole numbers in base ten for young children. But, as you are about to discover, adults can become confused by the information they have already acquired with the base-ten system. We have attempted to minimize the chances of confusion by dealing totally with a base-five system. Do not just "convert to base ten" to do the exercises in this chapter because that would defeat the entire purpose of these activities.

We recommend that you keep coming back to the beginning of the chapter (or previous exercises) to maintain a sense of continuity in the system we are constructing step by step. The first section develops a naming system for whole numbers. Thereafter the emphasis is on the basic operations of addition, subtraction, multiplication, and division with their respective properties.

We cannot overemphasize the importance of *analysis* and *reflection* on your part as you do this chapter. Base-five numeration is *not* the important aspect of the chapter. It is the *system* of naming numbers and the *properties* that hold for the operations on whole numbers that we are getting at. Our experience with this approach has led us to believe you can learn a great deal about the naming of numbers and basic properties of numbers in an enjoyable atmosphere through studying this base-five system.

1. *Naming numbers*

For this entire chapter, consider the green strip the basic unit. This strip will represent the number 1. Each of the other strips will represent corresponding numbers, as shown in Figure 1.

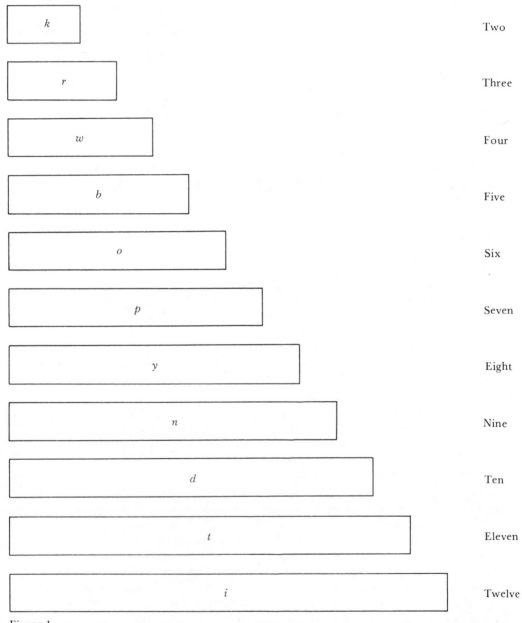

k	Two
r	Three
w	Four
b	Five
o	Six
p	Seven
y	Eight
n	Nine
d	Ten
t	Eleven
i	Twelve

Figure 1

The number 13 can be represented by a chain of a pink and a green. The number 5 can be represented by these chains:

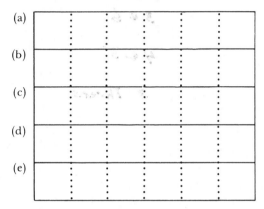

(a) | b
(b) | g | w
(c) | k | r
(d) | r | k
(e) | g | k | k

1. Mark the array to show various chains representing the number 6.

(a)
(b)
(c)
(d)
(e)

2. Use your strips to determine which number the rows of each array represent.

(a)

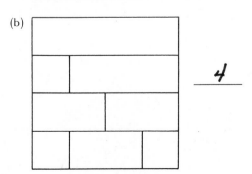

$\underline{3}$

(b)

$\underline{4}$

3. Name the individual strips by number.

EXAMPLES

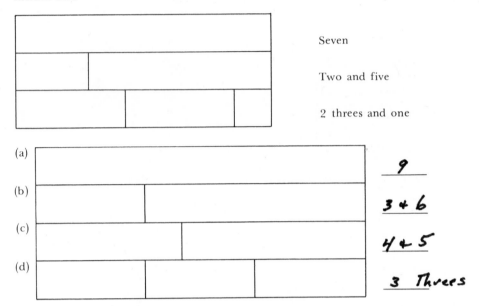

Seven

Two and five

2 threes and one

(a)   *9*

(b)   *3 + 6*

(c)   *4 + 5*

(d)   *3 Threes*

4. Under the sytem we have outlined, the number 1 can be represented in only one way using a chain: as one unit strip. The number 2 can be represented in just two ways using chains: as a black strip or as two unit strips. The number 3 can be represented in four ways since a chain of a green and a black is different from a chain of a black and a green.

   a. How many ways can the number 4 be represented as a chain of strips? *8*

   b. How many ways can the number 5 be represented as a chain of strips? *16*

   c. Fill in Table 1 and make a conjecture about the number of ways any given number can be represented by a chain of strips.

Table 1   Strip combinations

Number	One	Two	Three	Four	Five	Six	Seven	Eight
Ways to represent	*1*	*2*	*4*	*8*	*16*	*32*	*64*	*128*

5. Give the single-number name to these chains.

EXAMPLE

Fifteen

(a)

_13_

(b)

_14_

6. Another way to name the number represented is to record which colors were used. For example, (a) of problem 5 could be called purple and orange instead of thirteen. And (b) of problem 5 could be called two light blue and one white. Name the following chains in the same way:

(a)

2 Rose & 3 black

(b)

1 light blue + 4 bk

(c)

3 rose +1 white +1 green

(d)

3 white + 5 green

This is a cumbersome procedure for representing numbers because of the many ways each number can be represented. Given various combinations of colors, it is almost impossible to remember what number a particular combination stands for. To facilitate communication, we shall invent a system for using certain colors to name all numbers.

We shall arbitrarily select a particular strip to use as a standard measure, along with the green units. This way numbers will always be named in terms of the units and the specified color. The color we begin with will be light blue because we have four fingers and a thumb on our hands. (This will be useful for calculation later on.)

7. Determine how many light blue and unit strips are just as long as these chains.

EXAMPLE

Two light blue strips and four units

(a)

3 lT. blue & 1 unit

(b)

2 liT. bl + 2 units

(c)

3 light blue

(d)

2 light blue + 3 units

We call the shaded chain a *standard* light blue chain. We shall refer to this on many occasions.

8. For even greater convenience, use these marks when recording the number of strips in a standard light blue chain:

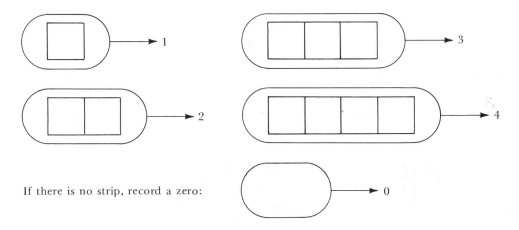

If there is no strip, record a zero: → 0

For combinations of strips that are longer than a light blue strip, record the number of light blue strips and the number of units. The rule for long chains is that we use as many light blue strips as possible before using unit strips. Accordingly, we shall not have five or more unit strips in a standard chain.

EXAMPLE

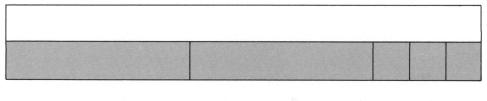

_____2_____ light blue strips and __3__ units

Fill in the blanks.

(a)

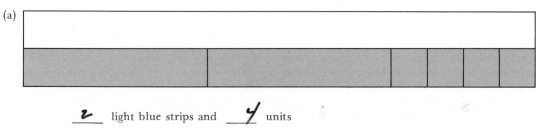

__*2*__ light blue strips and __*4*__ units

(b)

__*3*__ light blue strips and __*0*__ units

(c)

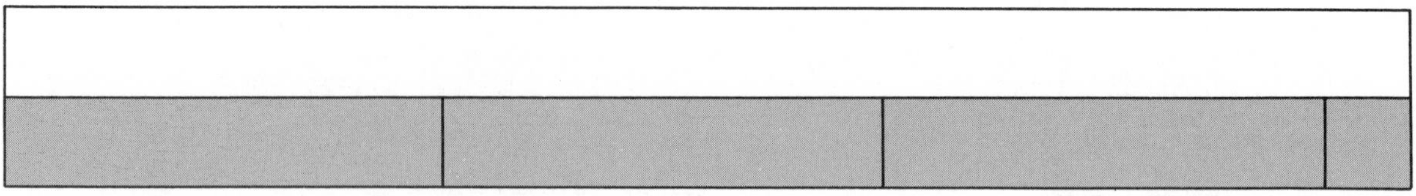

3 light blue strips and **1** units

(d)

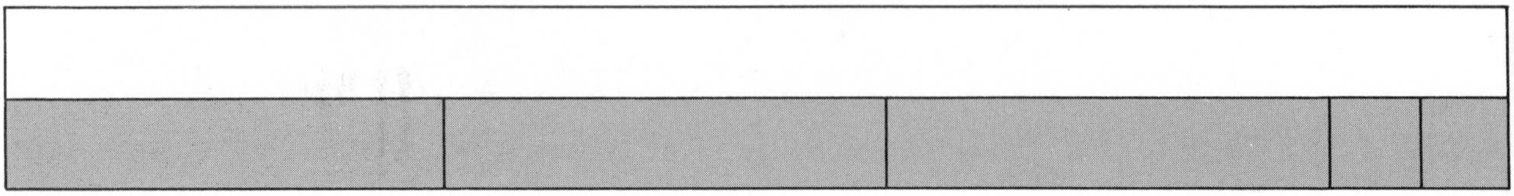

3 light blue strips and **2** units

9. Use your strips to determine the number of light blue strips and units for these chains. Rearrange the original chains if you wish.

(a)

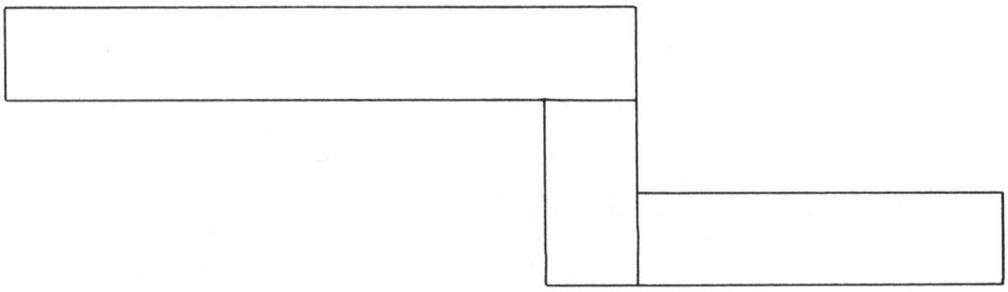

2 light blue strips **3** units

(b)

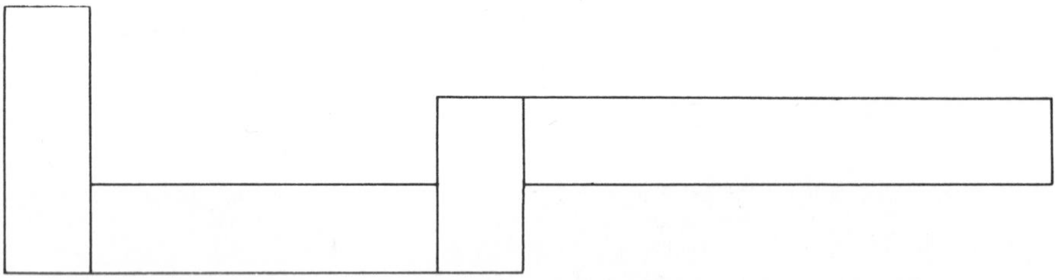

3 light blue strips **0** units

(c)

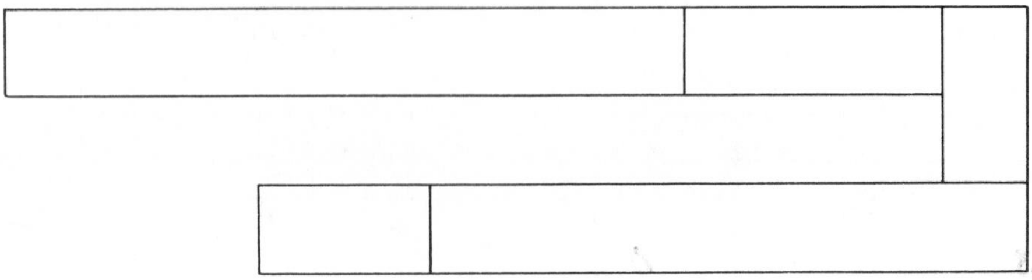

4 light blue strips **2** units

10. For representing exceptionally long chains, we find it convenient to collect the light blue strips in groups of five to form a light blue flat. The following collection of strips can be represented by one light blue flat, two light blue strips and one unit. (Check with your strips and flats.)

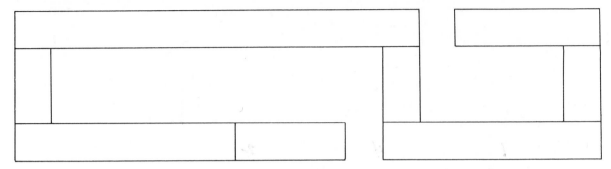

11. Represent these collections of strips with light blue flats, light blue strips, and units.

EXAMPLE

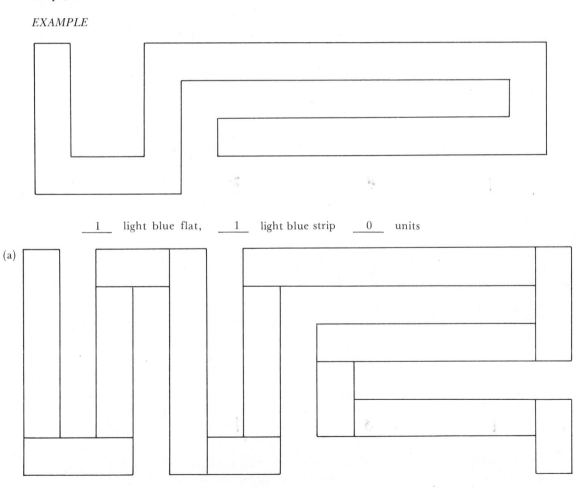

    1   light blue flat,   1   light blue strip   0   units

(a)

    2   light blue flats,   0   light blue strips   2   units

(b)

_____1_____ light blue flats, _____4_____ light blue strips, _____2_____ units.

(c)

_____1_____ light blue flats, _____4_____ light blue strips, _____2_____ units.

(d)

_____1_____ light blue flats, _____3_____ light blue strips, _____1_____ units.

To establish a means of verbally communicating these representations for numbers without confusing them with the names you already know, we shall arbitrarily name the light blue strip a *quint* and the light blue flat a *quare*. The next name would be *qube*. There is a one-to-five relationship between each collection. These names are sufficient for naming the numbers we shall deal with in this section. We now have a means of naming numbers with numerals and words based on collections of fives. This system is referred to as a base-five system of numeration.

12. Continue Table 2 with appropriate names and numerals.

Table 2   Numeral names in base five

Representation	Name	Numeral	
		quints	ones
	one		1
	two		2
	three		3
	four		4
	quint	1	0
	quint one	1	1
	quint two	1	2
	quint three	*1*	*3*
	quint four	*1*	*4*
	two quint	*2*	*0*
		2	*1*
		2	*2*
		2	*3*

As you continue this recording system for chains, you soon realize that it is cumbersome to place the light blue strips in a long line. Consequently, we arbitrarily represent a long chain in this way: Find how many quares, quints, and ones are in the standard chain for Figure 2.

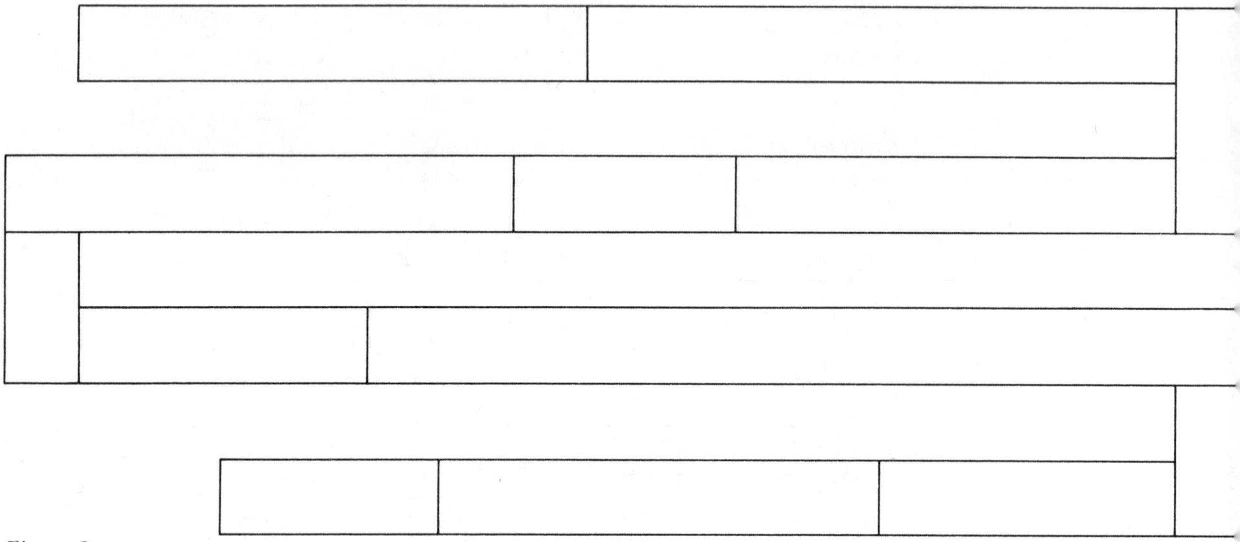

Figure 2

Now place the ones, quints, and quares in three piles from right to left as shown in Figure 3. Record numerals under each frame to indicate how many of each are in the frames.

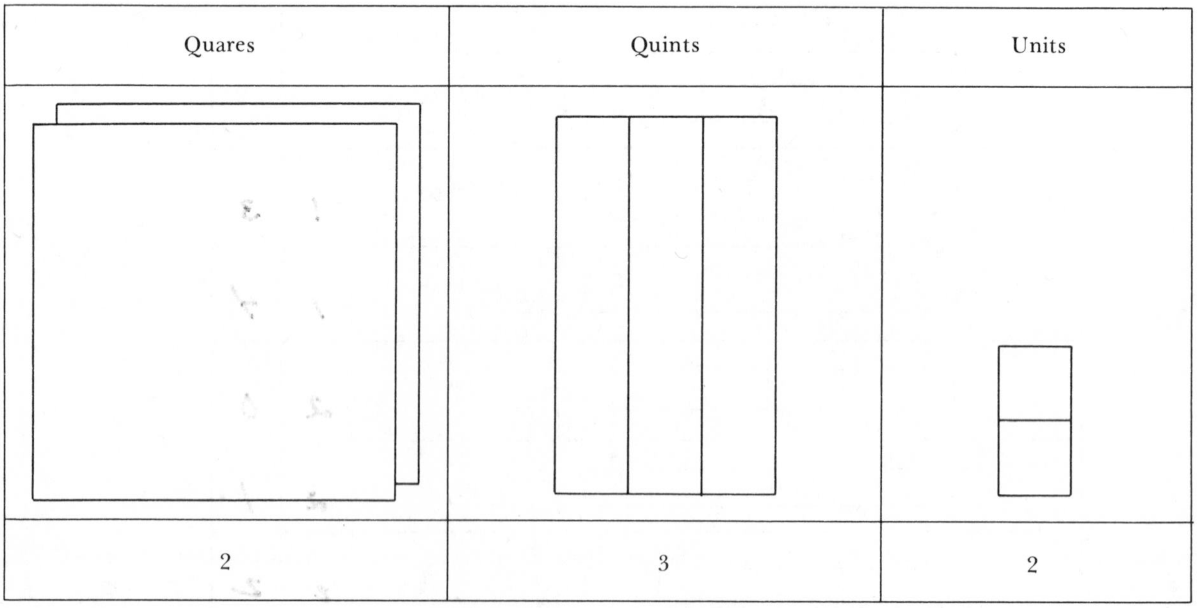

Quares	Quints	Units
2	3	2

Figure 3

This light blue standard would be called "2 quare, 3 quint, 2." Here are some more examples. Note the subscript *b* which indicates base five.

Quares	Quints	Ones	Name	Numeral
			Four quint four	44_b
			Quare	100_b
			Quare one	101_b
			Quare two	102_b
			Quare three	103_b
			Quare four	104_b
			Quare quint	110_b

13. Represent these collections with words and numerals of the base-five system.

(a)

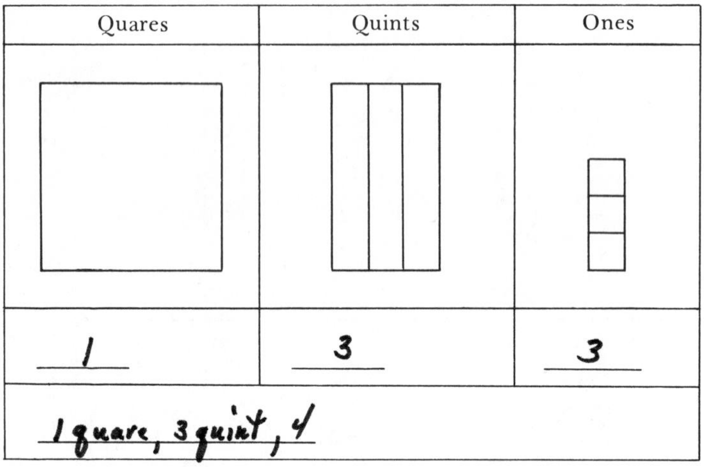

Quares	Quints	Ones
1	*3*	*3*
1 quare, 3 quint, 4		

(b)

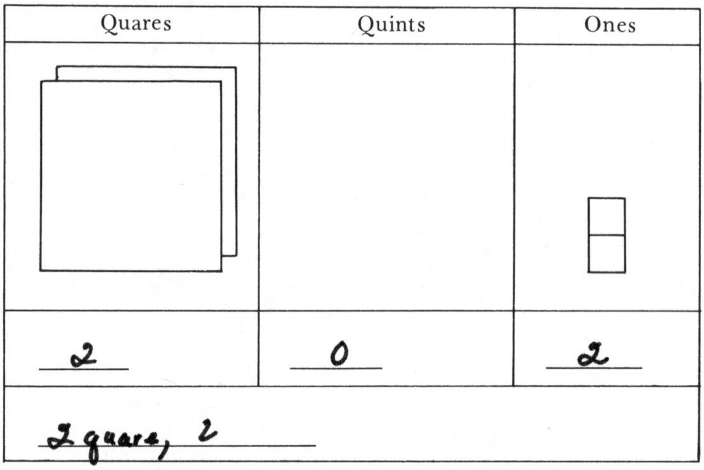

Quares	Quints	Ones
2	*0*	*2*
2 quare, 2		

(c)

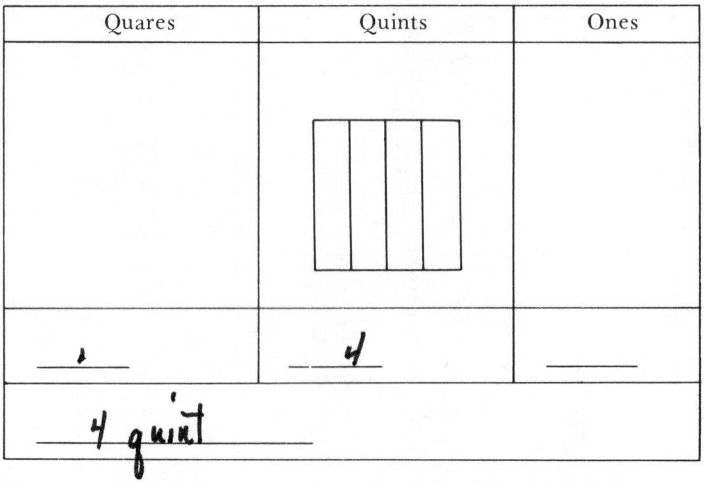

Quares	Quints	Ones
1	*4*	
4 quint		

(d)

Quares	Quints	Ones
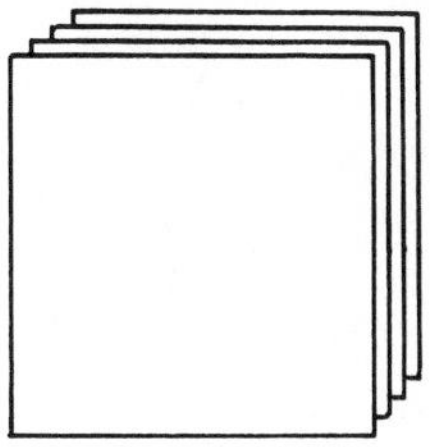	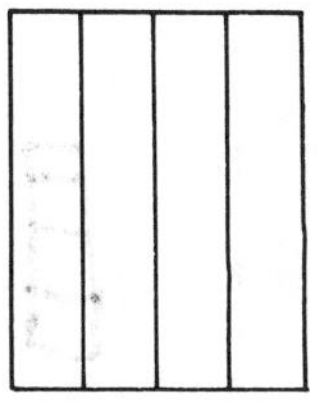	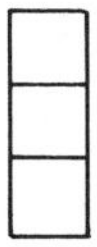
_____3_____	_____4_____	_____3_____
3 quare, 4 quint, 4		

14. Represent these numerals with sketches of units, light blue strips, and light blue flats:

(a)

Quares	Quints	Ones

_____	_____3_____	_____2_____
Three quint two		

(b)

Quares	Quints	Ones

_____1_____	_____1_____	_____1_____
Quare quint one		

(c)

Quares	Quints	Ones
4	0	4

Four quare four

(d)

Quares	Quints	Ones
2	0	0

Two quare

15. Devise a naming system for large numbers expressed in base-five notation. What would you call these numbers?

a. $2,314b$

b. $3,001b$

c. $4,440b$

d. $22,321b$

e. $120,321b$

f. $2,232,012b$

16. How are large numbers named in base-ten notation? What is the pattern? Is there a similar pattern to the system you described in problem 15? Can you devise a system for the base-five numerals that is analogous to our base-ten system?

2. Addition

In Section 1 we developed a system for naming numbers using the light blue quares, the light blue quints, and the green ones. This was done to establish a convenient means of communicating concepts of number and computational procedures. Nevertheless, we still have two ways of dealing with whole numbers: using a variety of colored strips for a particular number or using only light blue flats, light blue strips, and green units. There are times when one procedure has advantages over the other.

1. Select a white strip and purple strip. Lay them end to end. Name one strip which is as long as this chain.

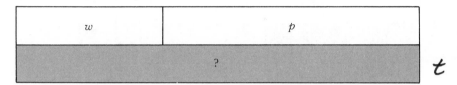

t

2. Find similar results for all possible pairs of strips from our basic set of 12. Complete Table 3.

Table 3   Letter addition

#	g	k	r	w	b	o	p	y	n	d	t	i
g	k	r	w	b	o	p	y	n	d	t	i	
k	r	w	b	o	p	y	n	d	t	i		
r	w	b	o	p	y	n	d	t	i			
w	b	o	p	y	n	d	t	i				
b	o	p	y	n	d	t	i					
o	p	y	n	d	t	i						
p	y	n	d	t	i							
y	n	d	t	i								
n	d	t	i									
d	t	i										
t	i											
i												

3. You probably had some doubts about what to place in the table for results that required a strip longer than pink. There are several ways you could have completed the table. One would have been to use the pink and whatever other strip it took to make the length. Another would have been to use our *standard chain* idea for all the lengths. This would have given these results:

EXAMPLE

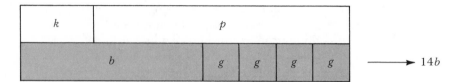

EXAMPLE

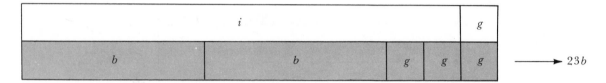

Now complete Table 4 using standard chain notation for the entire table.

Table 4   Addition using standard chains

+	1	2	3	4	10b	11b	12b	13b	14b	20b	21b	22b
1	2	3	4	10b	11b	12b	13b	14b	20b	21b	22b	23b
2	3	4	10b	11b	12	13	14	20	21	22	23	24
3	4	10b	11	12	13	14	20	21	22	23	24	30
4	10b	11	12	13	14	20	21	22	23	24	30	31
10b	11	12	13	14	20	21	22	23	24	30	31	32
11b	12	13	14	20	21	22	23	24	30	31	32	33
12b	13	14	20	21	22	23	24	30	31	32	33	34
13b	14	20	21	22	23	24	30	31	32	33	34	40
14b	20	21	22	23	24	30	31	32	33	34	40	41
20b	21	22	23	24	30	31	32	33	34	40	41	42
21b	22	23	24	30	31	32	33	34	40	41	42	43
22b	23	24	30	31	32	33	34	40	41	42	43	44

4. Since the end-to-end operation on strips is the same as addition of whole numbers, we replaced the end-to-end symbol ⊕ with a + sign. The table continues indefinitely to the right and down. Using this information, form a conjecture about whether addition on the set of whole numbers $\{0, 1, 2, 3, 4, 10b, 11b, 12b, \ldots\}$ has any of the following properties:

a. Closure: If a and b are any two whole numbers, then $a + b$ is a whole number.

yes

b. Associativity: If a, b, and c are any three whole numbers, then $(a + b) + c = a + (b + c)$.

yes

c. Additive identity element: There is a whole number 0 such that, when added to any whole number a, results in the number a; $a + 0 = a$.

yes

d. Additive inverses: For each whole number a there is a whole number b such that $a + b = 0$.

no

e. Commutativity: If a and b are whole numbers, then $a + b = b + a$.

yes

3. Subtraction

In Section 2 you took two strips, laid them end to end, and determined the resulting chain. The two strips represented *addends,* and the resulting chain represented the *sum.* To assist in communication about these resulting chains, we used the standard chain notation. In this section the starting point is one of the addend strips and the sum chain. Then the missing addend strip is determined. We shall still record numbers using the base-five system of standard chains in light blue.

EXAMPLE Find the missing link in the sum chain $22b$ if one of the addends is 4.

Answer: $13b$

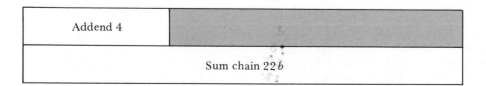

1. Now find the missing addends for these sum chains:

(a)
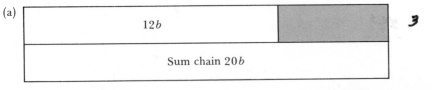
12b   **3**

Sum chain 20b

(b)
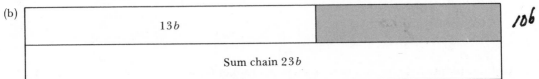
13b   *106*

Sum chain 23b

(c)
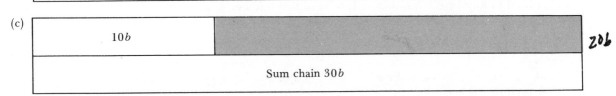
10b   *20b*

Sum chain 30b

(d)
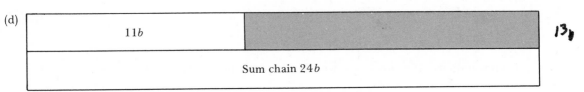
11b   *13ı*

Sum chain 24b

2. Find similar results for other sum chains and addends. Place the results in the table of missing addends. Represent missing addends as standard chains.

   a. What patterns can you see in Table 5?

   b. Why did you not complete the table?

   c. Can you make a much bigger table? How big could this table be?

3. Find the missing addends:

Sum chain	Given addend	Missing addend
a. 20b	12b	**3**
b. 22b	4	**13**
c. 34b	14b	**20**
d. 101b	34b	**12**

Table 5   Missing Addends

Given addend

—	0	1	2	3	4	10b	11b	12b	13b	14b	20b	21b	22b
0	0												
1	1	0											
2	2	1	0										
3	3	2	1	0									
4	4	3	2	1	0								
10b	10b	4	3	2	1	0							
11b	11b	10b	4	3	2	1	0						
12b	12b	11	10	4	3	2	1	0					
13b	13	12	11	10	4	3	2	1	0				
14b	14	13	12	11	10	4	3	2	1	0			
20b	20	14	13	12	11	10	4	3	2	1	0		
21b	21	20	14	13	12	11	10	4	3	2	1	0	
22b	22	21	20	14	13	12b	11	10	4	3	2	1	0

(Sum chain — row labels)

e. 221b   111b   *110*

f. 300b   24b   *221*

4. The process of finding the missing addend is analogous to subtraction. We can therefore write a subtraction sentence for each of the operations of problem 3. Complete these sentences:

	Sum chain		Given addend		Missing addend
a.	20b	—	12b	=	~~12~~ *3*
b.	22b	—	4	=	~~11~~ *13*
c.	34b	—	14b	=	~~14~~ *20*
d.	101b	—	34b	=	~~43~~ *12*
e.	221b	—	111b	=	~~24~~ *110*
f.	300b	—	24b	=	~~105~~ *221*

5. Write subtraction sentences and then find the missing addend.

	Sum chain	Given addend	Subtraction sentence
a.	24b	12b	$24b - 12b = 12b$
b.	30b	14b	$30 - 14 = 11$
c.	34b	20b	$34 - 20 = 14$
d.	120b	22b	$120 - 22 = 73$
e.	104b	30b	$104 - 30 = 24$
f.	242b	104b	$242 - 104 = 133$

6. Since finding the missing addend is the same as subtraction, we can use the − sign and study the properties of subtraction on whole numbers. Make conjectures about each of the following for subtraction on the set of whole numbers:

a. Closure: $a - b$ is a whole number.

no

b. Associativity: $(a - b) - c = a - (b - c)$.

no

c. Subtractive identity element: $a - \square = a = \square - a$.

no

d. Subtractive inverses: $a - b = \square$.

no

e. Commutativity: $a - b = b - a$.

no

7. Perhaps you are a little insecure after the last exercise since subtraction is a difficult operation to analyze on the set of whole numbers. In the first place, there is no closure because there are no whole number solutions to equations such as

$23b - 34b = \square$
$14b - 22b = \square$

Additive inverse

Based on your past experience with arithmetic systems, suggest an arrangement that would allow subtraction of *any* two whole numbers. Can the strips represent this new system?

no

4. *Multiplication*

As previously stated, there are two ways we can represent whole numbers: with a combination of various colored strips or with a standard color strip. The standard chain communicated easily when everyone used it to express whole numbers. We shall follow the rule of always expressing the end results in standard chain notation. For example, take a white strip and an orange and lay them in an arrangement like Figure 4. Fill in the shaded region with orange.

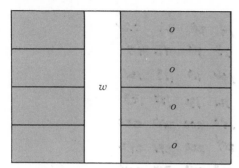

Figure 4

Discard the white strip. Lay the four oranges end to end to make a chain, as shown in Figure 5. Make a standard chain underneath. Record the numeral.

Figure 5

Another way would be to represent the orange and white strips with standard chains at the outset. Then the figures would look like Figure 6.

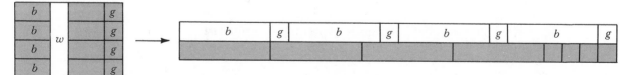

Figure 6

From a manipulative standpoint, using standard chains initially can become rather cumbersome. The small units are hard to manage, and with larger numbers there are many strips. But the manipulation of standard chains is directly analogous to calculating procedures we shall learn later on. Look carefully at the stacking procedure described for the chains $12b$ and $23b$. Do you see a relationship to a calculating procedure for:

$$23b$$
$$\times\,12b$$

1. Complete the multiplication in Table 6 using the following procedure:

   a. Take two colored strips and make an array such as in Figure 4.

   b. Stretch them out and find the standard chain such as in Figure 5.

   c. Record the result using standard chain notation.

Table 6   Multiplication table

X	1	2	3	4	10b	11b	12b	13b	14b	20b	21b	22b
1	1	2	3	4	10	11	12	13	14	20	21	22
2	2	4	11b	13	20	22	24	31	33	40	42	44
3	3	11b	14	22	30	33	41	44	102	110	113	121
4	4	13	22	31	40	44	103	112	121	130	134	143
10b	10	20	30	40	100	110	120	130	140	200	210	220
11b	11	22	33	44	110	121	132	143	204	220	231	242
12b	12	24	41	103	120	132	144	211b	223	240	302	314
13b	13	31	44	112	130	143	211	224	242	310	323	341
14b	14	33	102	121	140	204	223	242	311	330	344	413
20b	20	40	110	130	200	220	240	310	330	400	420	440
21b	21	42	113	134	210	231	302	323	344b	420	441	1012
22b	22	44	121	143	220	242	314	341	413	440	1012	1034

   d. What patterns do you see in the table?

   symmetry on diagonal

   e. What shortcuts did you find for speeding up your task?

   f. Can you use a quick procedure for determining products such as:

   $$\begin{array}{r} 21b \\ \times\ 14b \\ \hline 134b \\ 210b \\ \hline 344b \end{array}$$

2. Make a stack of a black, rose, and white (Figure 7).

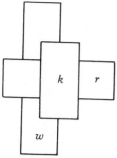

Figure 7

Now make a floor under the black with another rose strip (Figure 8).

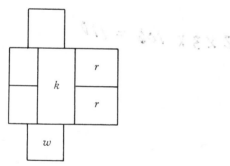

Figure 8

Discard the black, and stretch out the two roses into a chain (Figure 9).

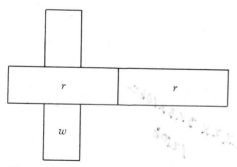

Figure 9

Now make a white floor under the two rose strips (Figure 10). Discard the two roses and find the standard chain for the white strips.

w	w	w	w	w	w
r			r		

Figure 10

You should have a standard chain of 44b. This represents $2 \times 3 \times 4 = 44b$.
Use the same procedure on these stacks to find the standard chain:

(a)

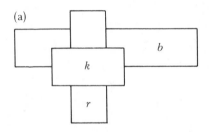

$2 \times 3 \times 10b = 110$

(b)

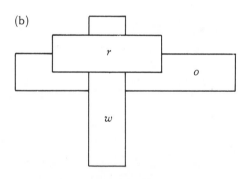

$3 \times 4 \times 11b = 242b$

(c)

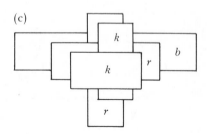

$2 \times 2 \times 3 \times 3 \times 10b = 1210b$

(d)

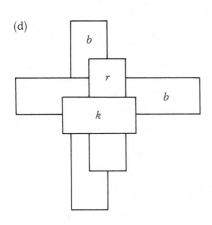

$2 \times 3 \times 10b \times 10b = 1100b$

3. Represent these numbers with strips, make a stack with them, find the standard chain, and then write the resulting base-five numeral.

EXAMPLE   2, 2, 10*b*   <u>40*b*</u>

a. 3, 4, 10*b*   <u>22*b*</u>

b. 4, 11*b*, 11*b*   <u>103*4*</u>

c. 2, 3, 10*b*, 12*b*   <u>1320</u>

d. 3, 3, 3, 3   <u>311</u>

e. 10*b*, 10*b*, 3   <u>300</u>

f. 10*b*, 10*b*, 10*b*   <u>1000</u>

g. 2, 10*b*, 2, 10*b*   <u>400</u>

4. Check the operation of multiplication on the set of whole numbers $\{0, 1, 2, 3, 4, 10b, 11b, \ldots\}$ for each of these properties:

  a. Closure   *yes*

  b. Associativity   *yes*

  c. Existence of an identity element   *yes 1*

d. Existence of an inverse for each element

e. Commutativity

yes

5. *Division*

Make a chain that is represented by $40b$.

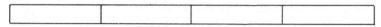

Figure 11

Our objective is to represent this chain with a *stack* of strips that includes a specific strip. Here is an example with white and the $40b$ of Figure 11.

a. Lay out white strips along the $40b$ chain as in Figure 12.

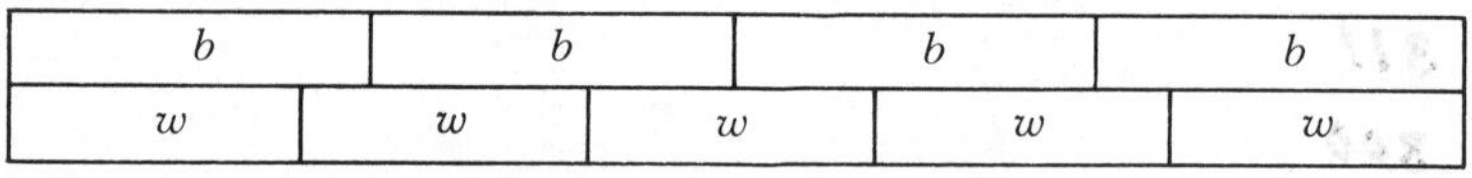

b	b	b	b
w	w	w	w	w

Figure 12

b. Arrange the white strips of $40b$ so that they form a rectangular array that is light blue high (Figure 13). Use as many strips as possible.

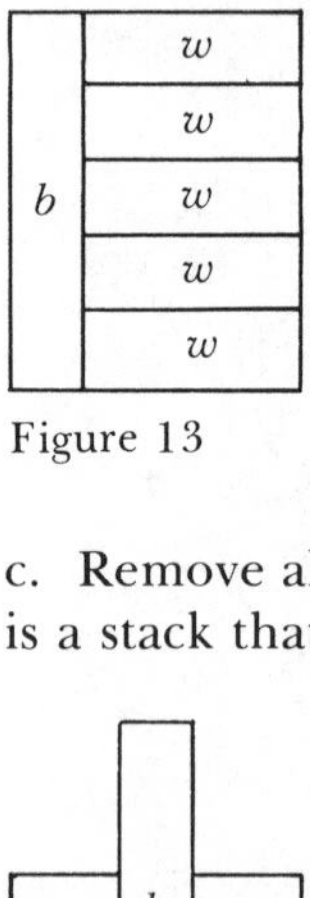

b	w
	w
	w
	w
	w

Figure 13

c. Remove all but one white and make the stack (Figure 14). The result is a stack that includes a white strip. The "other" strip is light blue.

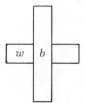

Figure 14

d. Consequently $40b \div 4 = 10b$

1. Make two-strip stacks with the specified strip for each of the numbers given.

EXAMPLE   Make a two-strip stack that includes a black strip for 13b.

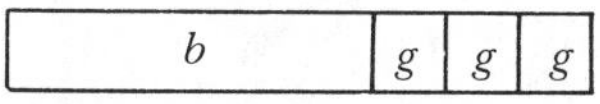

| | | b | | | g | g | g |

Figure 15

Given the chain in Figure 15, we must successively lay out black strips as in Figure 16. Now arrange the black strips as in Figure 17. The other strip of the stack is white so that the result of 13$b \div 2 = 4$.

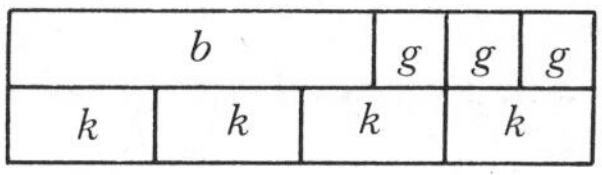

| | b | | | g | g | g |
| k | | k | k | | | k |

Figure 16

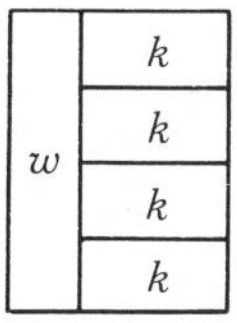

Figure 17

   a. Make a two-strip stack that includes an orange for 22b. The other strip is: **black**

   b. Make a two-strip stack that includes an orange for 44b. The other strip is: **white**

   c. Make a two-strip stack that includes a light blue for 110b. The other strip is: **orange**

   d. Make a two-strip stack that includes a white for 31b. The other strip is: **white**

2. Make stacks with as many nongreen as possible for these numbers.

EXAMPLE 1   <u>31b</u>   The most strips possible for this number is four blacks.

EXAMPLE 2   <u>103b</u>   The most strips possible for this number is three: two blacks and one purple.

EXAMPLE 3   <u>41b</u>   Rose and purple.

EXAMPLE 4   <u>34b</u>   *no stack possible.*

EXAMPLE 5   <u>110b</u> k, r, and b.

   a. <u>22b</u>   **2 blacks + 1 rose**

   b. <u>33b</u>   **one black + 2 rose**

   c. <u>121b</u>   **2 blacks + 2 rose**

   d. <u>124b</u>   **1 rose,   2 lite blue + 1 rose**

3. Complete Table 7 for this operation of finding the other strip of a stack. Select a number from the left column. Now represent this number with a two-strip stack that includes a number from the upper row. The other strip of the stack should be recorded in the table.

EXAMPLE   The number $11b$ can be represented by a stack that includes a black strip (2). The other strip of the stack is a rose (3).

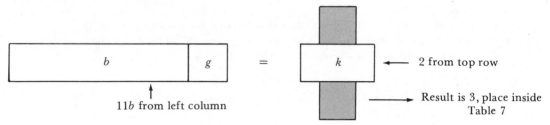

Figure 18

Table 7   Two-strip stacks

Make a two-strip stack including this number (but no ones).

Total stack represents this number	1	2	3	4	10b	11b	12b	13b	14b	20b	21b	22b
1	1	no										
2	2	no										
3	3	no	1									
4	4	2		1								
10b	10	no			1							
11b	11	3	2			1						
12b	12	no					1					
13b	13	4		2				1				
14b	14	no	3						1			
20b	20	10b			2					1		
21b	21	no									1	
22b	22	11	4	3		2						1

a. Not all combinations of numbers from the table can satisfy the requirements of this operation. Make a conjecture about the results of those you could complete.

Larger ÷ smaller

b. Make a conjecture about the numbers from the left column and top row that were the same such as 2 and 2, 13b and 13b, etc. What should the result be? Why?

$= 1$

c. Make a conjecture about pairs of numbers such as 4 and 12b where the second number is larger than the first number.

out of system

d. Make a conjecture about pairs of numbers such as 14b and 2 where it is impossible to use the 2 in a stack that represents 14b. Devise a way to handle this situation to come up with a result.

have remainders

6. *Properties of numbers*

Each of the exercises in this section deals with an important concept of arithmetic. You should do the activities that develop these notions carefully, recording observations for use in making various conjectures.

1. Make a chain that represents 11b. Can you make a chain of black strips that is just as long? (YES) NO (Circle one.) Try to make chains of black strips that are as long as chains that represent the numbers in Table 8.

a. Make a conjecture about the table you have just completed.
all numbers with yes circled are ÷ 2

b. Devise a way to tell whether or not a given number can be represented with an all-black chain. Describe it with examples.
sum of digits is even,

c. Which of these numbers can be represented by an all-black chain?

i. 34b *No*   ii. 101b *y*

iii. 2102b *No*   iv. 1111b *y*

v. 4444b *y*   vi. 10104b *y*

Table 8   Black chains

Number	Black chain as long
1	YES NO
2	✓ YES NO
3	YES NO
4	✓ YES NO
10*b*	YES NO
11*b*	✓ YES NO
12*b*	YES NO
13*b*	✓ YES NO
14*b*	YES NO
20*b*	✓ YES NO
21*b*	YES NO
22*b*	✓ YES NO
23*b*	YES NO
24*b*	✓ YES NO
30*b*	YES NO
31*b*	✓ YES NO
32*b*	YES NO

d. Numbers that can be represented by an all-black chain are called *even*. Numbers that cannot be represented by an all-black chain are called *odd*. Devise a rule that identifies odd and even numbers represented with base-five numerals and uses the digits of the numeral to do so.

sum of digits = even

2. Make a chain that represents 13*b*. Can you make a one-color chain that is just as long? (Do not use one single strip or any of the green units.)

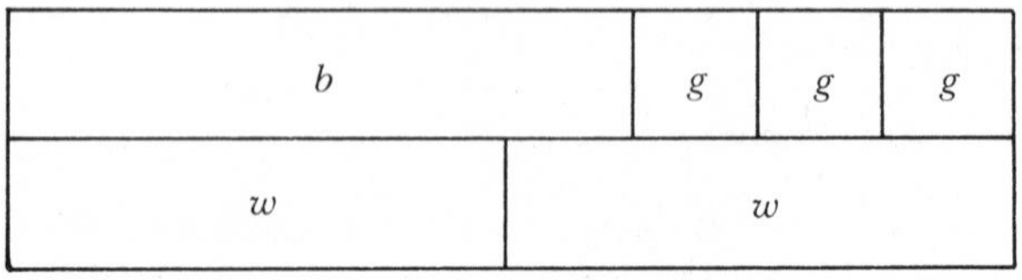

b			*g*	*g*	*g*
w			*w*		

White

Figure 19

The answer to the question is *yes*. Figure 19 is a sketch of one possible solution. There is another.

Find a one-color chain (not just units and more than one strip) for each number:

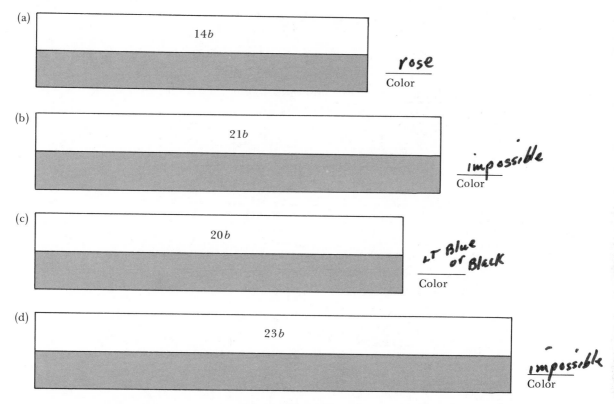

(a)

14*b*

rose
Color

(b)

21*b*

impossible
Color

(c)

20*b*

LT Blue or Black
Color

(d)

23*b*

impossible
Color

a. You can find one-color chains for some numbers (such as 14*b* and 20*b*) but not for others (such as 21*b* and 23*b*). Circle all the numbers in Table 9 for which you *cannot* find one-color chains.

Table 9   Primes

①	11*b*	21*b*	31*b*	41*b*	101*b*	111*b*
②	12*b*	22*b*	32*b*	42*b*	102*b*	112*b*
③	13*b*	23*b*	33*b*	43*b*	103*b*	113*b*
4	14*b*	24*b*	34*b*	44*b*	104*b*	114*b*
10*b*	20*b*	30*b*	40*b*	100*b*	110*b*	120*b*

The circled numbers (except the number 1) are called *primes*. The number 1 is not considered a prime by convention. Nonprime numbers are called *composites*.

b. A long time ago a famous mathematician made a conjecture about whole numbers and primes. It goes like this (somewhat paraphrased): Every nonprime number greater than one can be represented as a stack of primes. Demonstrate that this conjecture is true for these numbers:

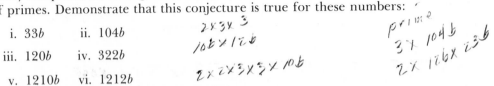

i. 33*b*   ii. 104*b*

iii. 120*b*   iv. 322*b*

v. 1210*b*   vi. 1212*b*

3. Make two chains, one of black and one of light blue, until they are the same length. Use as few strips as possible.

2	2	2	2	2

10*b*	10*b*

lcm = 20*b*

Figure 20

Each of the chains in Figure 20 represents the number 20*b*. This number is called the *least common multiple* (lcm) of 2 and 10*b*. Find the least common multiple of these pairs of numbers:

(a)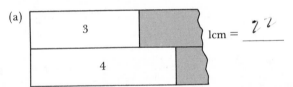

lcm = _____

(b)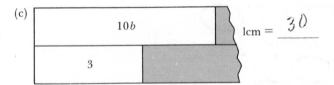

lcm = _____

(c)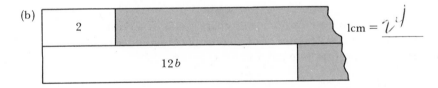

lcm = _____

(d)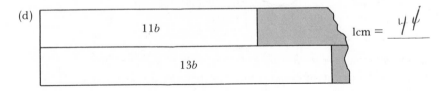

lcm = _____

Find the least common multiple for these pairs of numbers:

i. 13b, 22b lcm ___44b___ ii. 4, 13b lcm ___13b___

iii. 11b, 14b lcm ___33b___ iv. 20b, 30b lcm ___11$6$$b$___

v. 20b, 41b lcm ___13 2$C$$b$___ vi. 100$b$, 10$b$ lcm ___10$C$$b$___

4. Make chains for 22b and 30b. Using only one color for both, fill in the shaded regions below the two chains.

Rose is the only color that can be used for both chains.

a. Use a similar procedure for the pair of chains that follows. Name the color used.

Name of color ___*rose*___

b. Find a common color for making chains as long as these:

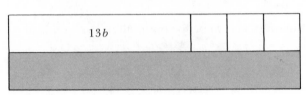

Name of color _whi̇te_

You may have used black for each of these chains. You could have also used white. For this exercise it is important that we use the white because they are longer than the black. The objective is to find the *greatest common color* (gcc) in making chains just as long as two or more number chains.

c. Find the greatest common color for making chains as long as these:

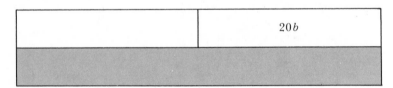

Name of color _Light blue_

d. Find the greatest common color for making chains as long as these pairs of numbers:

    i. $44b$ and $31b$ gcc _yellow_

    ii. $40b$ and $31b$ gcc _white_

    iii. $33b$ and $22b$ gcc _orange_

    iv. $121b$ and $44b$ gcc _pink_

e. For larger numbers the greatest common color designation is inappropriate. A *combination of strips* as a "common color" might be necessary. Consequently, a more general approach would be to identify the *common length* with a numeral. The name would then be *greatest common factor* (gcf). Find the greatest common factor for these pairs of numbers.

EXAMPLE   40*b* and 44*b*   gcf ___4___

    i. 110*b* and 140*b*   gcf _30 b_

    ii. 44*b* and 121*b*   gcf _22 b_

    iii. 14*b* and 33*b*   gcf _14 b_

    iv. 32*b* and 201*b*   gcf _32 b_

5. Select a divisor from the top row of Table 10. Check to see if it evenly divides the numbers in the dividend column. Place a check (√) in the appropriate squares to indicate that the first number you chose is a *factor* of the second number.

EXAMPLE   Select the divisor 12*b*. Since 12*b* is certainly not a factor of 4, 10*b*, or 11*b*, there will be no checks in the first three boxes. But 12*b* *is* a factor (along with 1) of 12*b* so place a check in that box. The next check would appear across from 24*b*.

    a. Devise a system for predicting when a number, expressed as a base-five numeral, is divisible by 2.

        Sum of digits   2 or 4   ÷ by 4

    b. Devise a system for predicting when a number, expressed as a base-five numeral, is divisible by 3.

        add   alternates — alternates =
               multiple of 3.

    c. Devise a system for predicting divisibility of numbers, expressed in the base five, for:

      i. 4   Sum   ÷ 4

      ii. 10*b*   Last digit = 0

iii. 11*b*

; 2 and 3

iv. 12*b*

easier to divide.

v. 13*b*

vi. 14*b*

No solution

vii. 20*b*

units place = 0
sum of other digits
even

viii. 21*b*

None

None

ix. 22*b*

Table 10   Quotients

Divisor

÷	2	3	4	10b	11b	12b	13b	14b	20b	21b	22b
4	✓		✓			No					
10b				✓		No					
11b	✓	✓				No					
12b						✓					
13b	✓		✓			No	✓				
14b		✓				No		✓			
20b	✓		✓			No			✓		
21b						No				✓	
22b	✓	✓	✓		✓	No					✓
23b						No					
24b	✓					✓					
30b		✓	✓			No					
31b	✓		✓			No	✓				
32b						No					
33b	✓	✓			✓	No		✓			
34b						No					
40b	✓		✓	✓		No			✓		
41b		✓				✓					
42b	✓					No				✓	
43b						No					
44b	✓	✓	✓		✓	No	✓				
100b				✓		No					
101b	✓					No					
102b		✓				No		✓			

Dividend

Summary

Successfully completing this chapter should give you a thorough background for understanding the nature of a number system and notation procedures. All the words we use to describe elements and operations of mathematics are so commonplace that we often do not recognize the complexity of the concepts or procedures they name.

Perhaps the most useful outcome you could achieve at this point is to realize that you can be in command of mathematics when you understand that it is a way of thinking and organizing. You should make mathematics work for you. It is logical, so you must be logical. It is complex and orderly, so you must be thorough and neat. It is precise, so you must learn to be careful.

For adults, the chapter you just completed should be most helpful in making numeration systems and properties meaningful. Most of us learned much of our mathematics in a meaningless, rote fashion. Rehearsing the same mathematical words and symbols to develop further understanding is difficult. Changing to another notation system helps us consider the ideas in a new, unfamiliar light. At the very least, the experience should give you some idea of the difficulties children can experience when learning school mathematics.

Of course, there is no need to go to this extreme with children if they first encounter number systems through a meaningful, structural model. Then mathematics is not some strange magic, but a logical system that is useful and easy to remember.

References

Hoffer, Alan R. "What You Always Wanted to Know about Six but Have Been Afraid to Ask." *The Arithmetic Teacher,* 20 (March 1973), 173–180.

King, Irv. "Giving Meaning to the Addition Algorithm." *The Arithmetic Teacher,* 19 (May 1972), 345–348.

Rahmlow, Harold. "Understanding Different Number Bases." *The Arithmetic Teacher,* 12 (May 1965), 339–340.

Ranucci, Ernest R. "Tantalizing Ternary." *The Arithmetic Teacher,* 15 (December 1968), 718–722.

Smith, Karl. "Inventing a Numeration System." *The Arithmetic Teacher,* 20 (November 1973), 550–553.

Vitt, Eddie E. "An Additive Numeral System Related to Place Value." *The Arithmetic Teacher,* 12 (March 1965), 212–215.

7

Base-ten arithmetic

In the previous chapter you spent a great deal of time on a system of representing whole numbers that is not used by anyone except some elementary school students, their parents, and their teachers. You were instructed in that system for exactly the same reason as the elementary school children: It confronts you with the basic concepts and procedures of a base and place-value numeration system. Your experiences with the base-five system will allow you to delve into the inner workings of the more common base-ten system.

Whenever small numbers are used, a base-five system is adequate for representing numbers and doing calculations. For larger numbers a numeration system with a larger base is more efficient. To create a better system, we now change our rules by redefining a *standard chain* as a chain made with dark blue strips and units, rather than light blue strips and units. Here are two examples for you to check to be sure you understand the change.

EXAMPLE 1

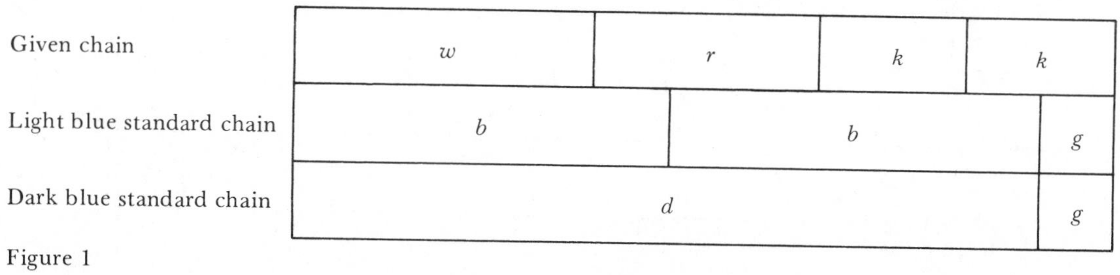

Given chain				
w		r	k	k

Light blue standard chain			
b		b	g

Dark blue standard chain		
d		g

Figure 1

The numeral for the dark blue standard chain in Figure 1 is 10. Since we shall use this naming system from now on, we shall not use a letter *d* after the numerals. Just remember that from now on *all* numerals you see are dark blue *standard* (or base-ten).

Since the maximum number of units we deal with has increased to nine, this new standard chain will use several more digits. Consequently, we shall use these marks for collections of five, six, seven, eight, and nine:

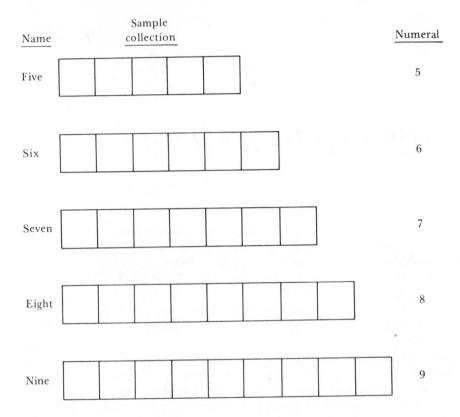

Name	Sample collection	Numeral
Five		5
Six		6
Seven		7
Eight		8
Nine		9

EXAMPLE 2   The dark blue standard for the chain in Figure 2 is 17.

y		o		r	
b	b		b	g	g
d		g	g g g	g	g

Figure 2

1. *Naming numbers*

For combinations of strips that are longer than a dark blue strip, record the number of dark blue strips and the number of unit strips. The rule for long chains is that we use as many dark blue strips as possible before using unit strips. Therefore we shall not have ten or more unit strips in a standard chain.

EXAMPLE

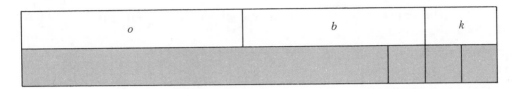

  1   dark blue strips and   3   unit strips

1. Fill in the blanks.

(a)

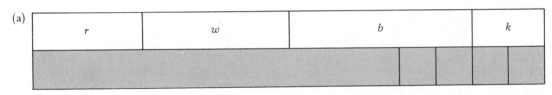

  1   dark blue strips and   4   unit strips

(b)

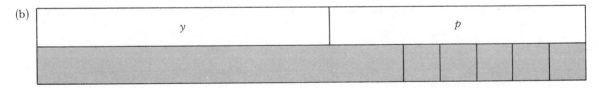

  1   dark blue strips and   5   unit strips

(c)

  1   dark blue strips and   6   unit strips

(d)

___1___ dark blue strips and ___7___ unit strips

2. Use your strips to determine the number of dark blue strips and units for the following chains:

(a)

___1___ dark blues ___2___ units

(b)

___1___ dark blues ___5___ units

(c)

___2___ dark blues ___2___ units

3. For exceptionally long chains of strips, we find it convenient to collect the dark blue strips in groups of ten to form a dark blue flat. The collection of strips in Figure 3 can be represented by one dark blue flat, one dark blue strip, and one unit. (Check with your flats, strips, and units.)

Figure 3

4. Represent the collection of strips in Figure 4 with dark blue flats, dark blue strips, and units.

EXAMPLE

$$\frac{0}{\text{Flats}} \qquad \frac{3}{\text{Strips}} \qquad \frac{0}{\text{Units}}$$

Figure 4

(a)

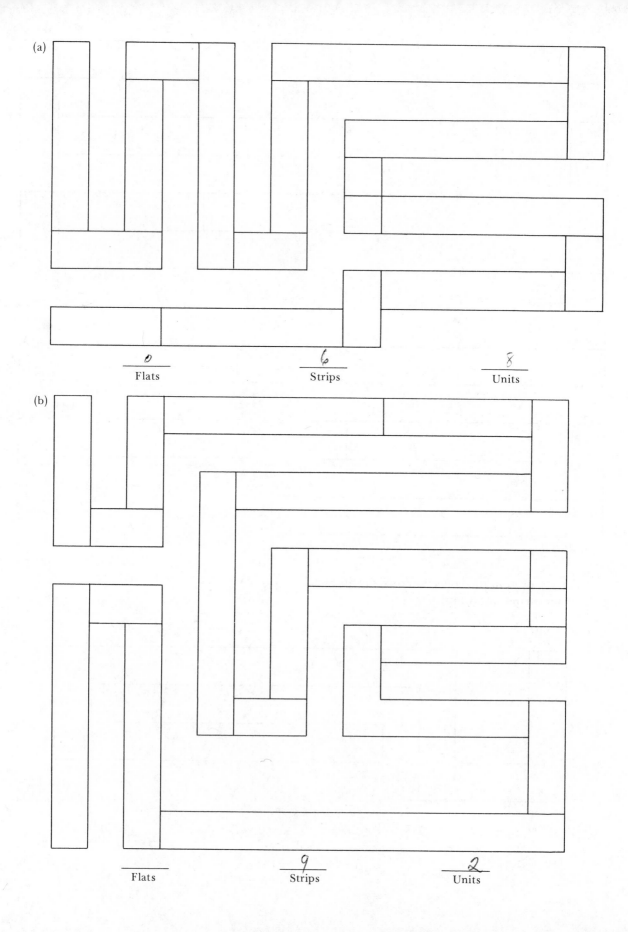

<u>0</u>
Flats

<u>6</u>
Strips

<u>8</u>
Units

(b)

<u>9</u>
Strips

Flats

<u>2</u>
Units

(c)

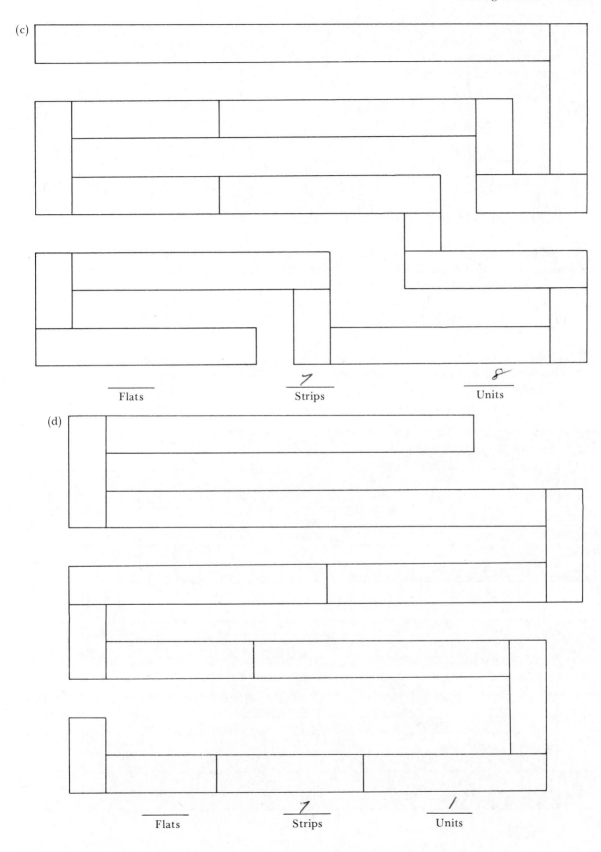

Flats   *7*   *8*
              Strips   Units

(d)

Flats   *7*   *1*
              Strips   Units

5. To establish a means of verbally communicating these representations for numbers, we shall arbitrarily name the unit square a *one,* the dark blue strip a *ten,* and the dark blue flat a *hundred* (Figure 5).

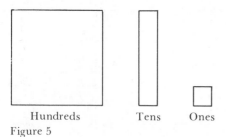

Hundreds   Tens   Ones
Figure 5

These shapes and names are sufficient for naming the numbers we shall deal with in this section. This gives us a means of naming numbers with *numerals* and *words* based on collections of tens.

6. Fill in the chart on the following page with appropriate figures, names, and numerals.

7. As you continue this recording system for strip chains, you soon realize that it is cumbersome to place the strips in a long line. Consequently, we arbitrarily represent a long chain in this way: Find how many hundreds, tens, and ones there are in the *standard* chain for the chain in Figure 6.

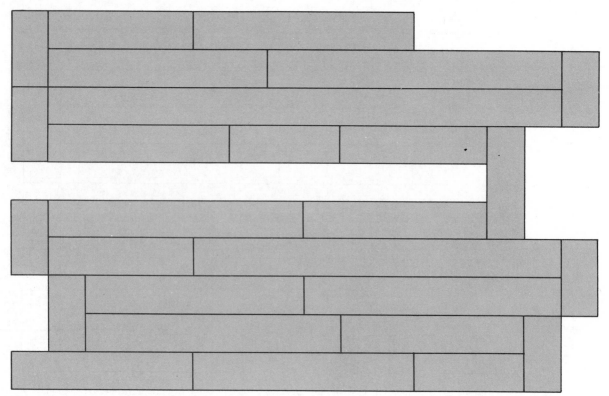

Figure 6

Representation	Name	Numeral
Example	Twelve	12
(a)		*9*
(b)		*10*
(c)		*11*
(d)		23
(e)	Thirty	*30*
(f)		100

Now place the ones, tens, and hundreds in three piles from right to left as shown in Figure 7. Record numerals under each collection to indicate how many of each are in the frames. This dark blue standard would be called "one hundred thirty-two."

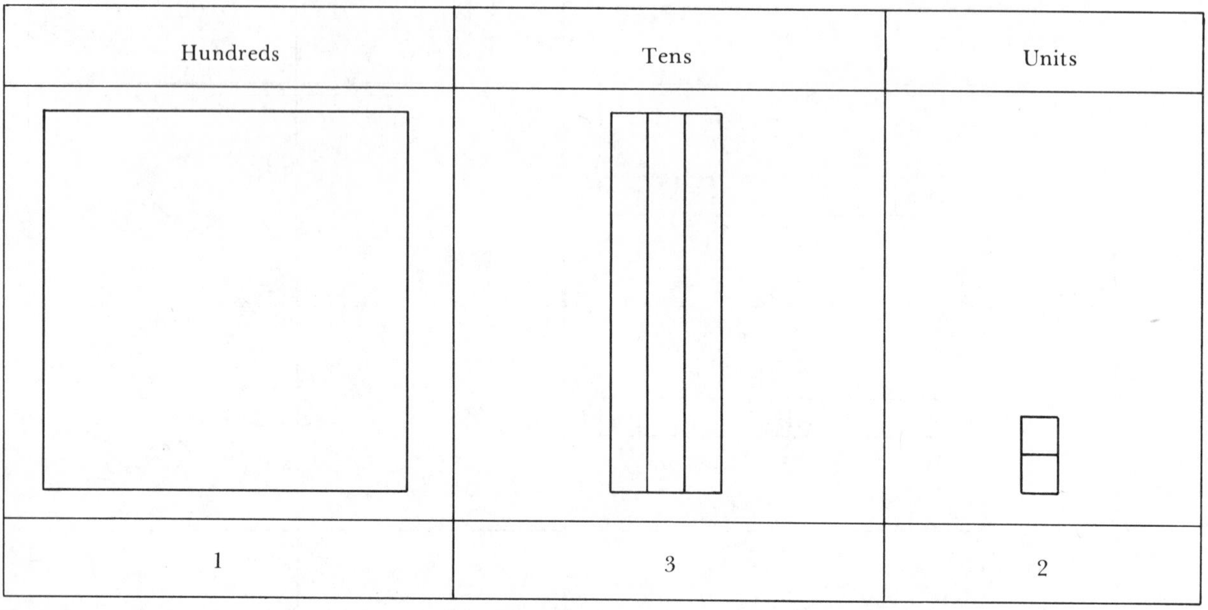

Hundreds	Tens	Units
1	3	2

Figure 7

8. Fill in the names and numerals for the collections of dark blue flats, dark blue strips, and units in the chart on the following page.

9. Represent the dark blue standards on page 140 with words and numerals as in Figure 8.

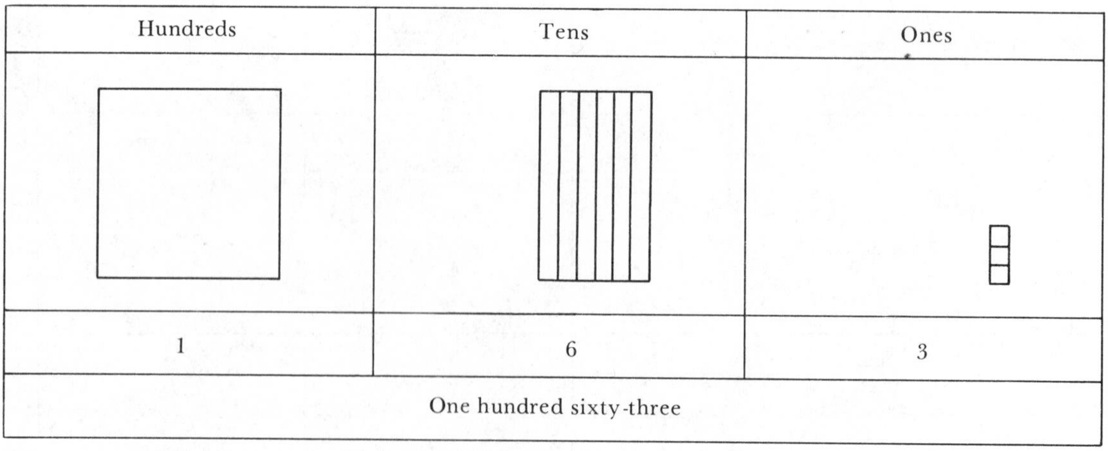

Hundreds	Tens	Ones
1	6	3
One hundred sixty-three		

Figure 8

	Hundreds	Tens	Ones	Name	Numeral
				Ninety-nine	99
(a)					100
(b)					102
(c)					134
(d)					110

(a)

Hundreds	Tens	Ones
2	0	2

(b)

Hundreds	Tens	Ones
	9	0

(c)

Hundreds	Tens	Ones
4	8	4

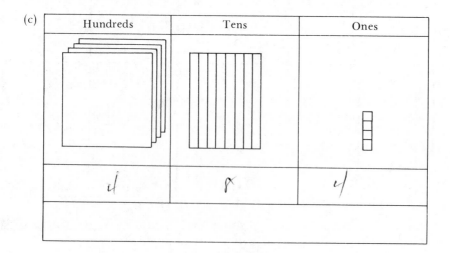

10. Represent these numbers with sketches showing the number of ones, tens, and hundreds. Write the numerals also.

   a. Thirty-one

Hundreds	Tens	Ones
	⫼	⨅

   b. One hundred eleven

Hundreds	Tens	Ones

   c. Four hundred four

Hundreds	Tens	Ones

d. Two hundred

Hundreds	Tens	Ones
.		

2. Addition

In Section 1 we developed a system for naming numbers using the dark blue flats (hundreds), the dark blue strips (tens), and the units (ones). Use this system to name the results of laying pairs of strips end to end.

EXAMPLE   Select a brown strip and rose strip. Lay them end to end. Name a standard chain that represents this chain (Figure 9).

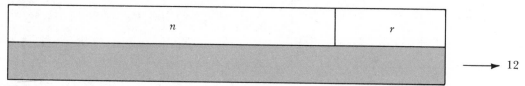

Figure 9

1. Find similar results for other combinations of strips, and complete Table 1.

2. Table 1 may be continued indefinitely to the right and down. Using this information, conjecture whether addition on the set of whole numbers {0, 1, 2, 3, 4, 5, 6, 7, . . .} has any of these properties:

a. Closure: If a and b are any two whole numbers, then $a + b$ is a whole number.

yes

Table 1   End-to-end results in dark blue standards

#	g	k	r	w	b	o	p	y	n	d	t	i
g	2	3	4									13
k	3	4										14
r	4											15
w												16
b												17
o						12						18
p												19
y												20
n												21
d												22
t												23
i											23	24

b. Associativity: If a, b, and c are any three whole numbers, then $(a + b) + c = a + (b + c)$.

yes

c. Additive identity element: There is a whole number 0 that, when added to any whole number a, results in the number a; $a + 0 = a = 0 + a$.

yes

d. Additive inverses: For each whole number a, there is a whole number b such that $a + b = 0$.

no

e. Commutativity: If a and b are whole numbers, then $a + b = b + a$.

yes

3. Subtraction

In Section 2 you took two strips, laid them end to end, and determined the resulting chain. The two strips represented addends, and the resulting chain represented the sum. To assist in communication about these resulting chains, the standard chain notation was used. In this section the starting point is one of the addend strips and the sum chain. The missing addend is to be determined. We shall record numbers using the base-ten system of the dark blue standard chain.

EXAMPLE   Find the missing addend in the sum chain 12 if one of the addends is 4. (See Figure 10.)

Answer: yellow

Addend w

Sum chain pink

Figure 10

1. Now find the missing addends for these sum chains:

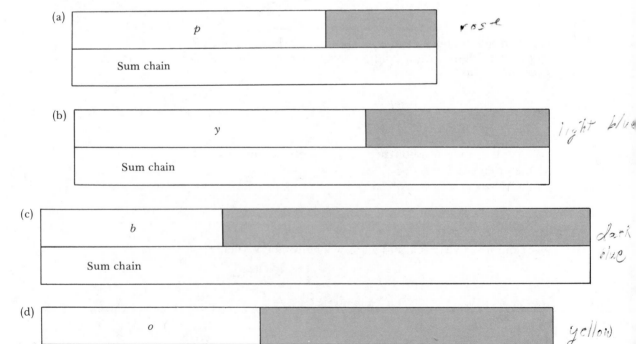

(a) p Sum chain   *rose*

(b) y Sum chain   *light blue*

(c) b Sum chain   *dark blue*

(d) o Sum chain   *yellow*

2. Find similar results for the other pairs of strips (sum chains and an addend), and place the results in the table of missing addends, Table 2. Represent all strips and chains as standard chains with base-ten numerals.

Table 2   Missing addends

Given addend

−	0	1	2	3	4	5	6	7	8	9	10	11	12
0	0												
1	1	0											
2	2	1	0										
3	3	2	1	0									
4			2		0								
5						0							
6							0						
7								0					
8	8	7	6	5	4	3	2	1	0				
9										0			
10											0		
11												0	
12							6						0

Sum chain

3. Find the missing addends. Use dark blue flats, dark blue strips, and units.

	Sum chain	Given addend	Missing addend
a.	20	8	12
b.	32	12	20
c.	57	24	33
d.	101	56	45
e.	200	132	68
f.	324	298	26

4. The process of finding the missing addend is analogous to subtraction. We can therefore write a subtraction sentence for each of the operations of problem 3. Complete these sentences:

	Sum chain		Given addend		Missing addend
a.	20	−	8	=	12
b.	32	−	12	=	20
c.	57	−	24	=	33
d.	101	−	56	=	45
e.	200	−	132	=	68
f.	324	−	298	=	24

5. Find the missing addend and then write subtraction sentences.

	Sum chain	Given addend	Subtraction sentence
a.	28	9	$28 - 9 = 19$
b.	35	14	
c.	59	37	
d.	81	46	
e.	162	75	
f.	200	142	

6. Since finding the missing addend is the same as subtraction, we can use the − sign and study the properties of subtraction on whole numbers. Make conjectures about each of the following for subtraction on the set of whole numbers.

a. Closure: $a - b$ is a whole number.

No

b. Associativity: $(a - b) - c = a - (b - c)$.

No

c. Subtractive identity element: $a - \Box = a = \Box - a$.

No

d. Subtractive inverses: $a - b = \Box$.

No   *no* *identity*

e. Commutativity: $a - b = b - a$.

No

7. As we discussed in the previous chapter, subtraction is a difficult operation to analyze on the set of whole numbers. In the first place, there is no closure because there are no whole number solutions to equations such as

$23 - 34 = \Box$
$14 - 22 = \Box$

Based on your past experience with arithmetic systems, suggest an arrangement that would allow subtraction of *any* two whole numbers. Can the strips represent this new system?

Additive inverse

No

4. *Multiplication*

Select a rose and a turquoise and stack them as in Figure 11.

Figure 11

Now fill in the shaded region with more *t* strips (Figure 12).

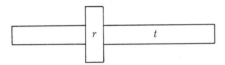

Figure 12

Discard the *r* strip and stretch out the *t*s into a chain (Figure 13).

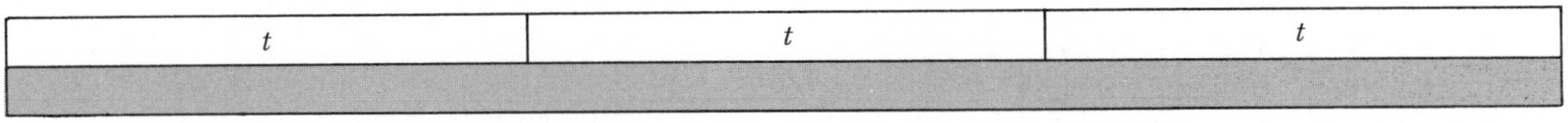

t	t	t

Figure 13

Under this chain fill in the shaded region to make a *d* standard chain. The numeral for this dark blue standard chain is 33.

Another procedure is to convert the *t* strip into a *d* standard before you begin the process of finding the result of stacking *t* and *r*. First convert *t* to *d* and *g* (Figure 14).

Now use *d*s and *g*s to fill in the shaded region (Figure 15).

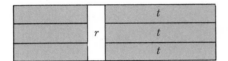

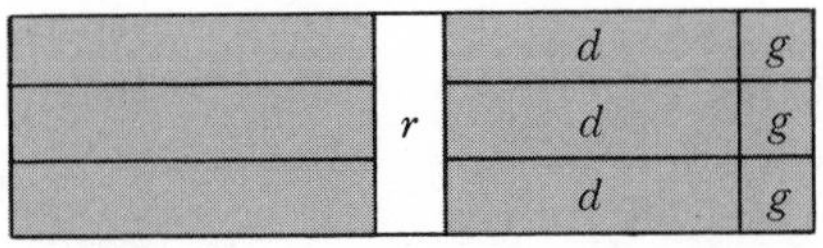

Figure 15

Discard the *r* strip and make a chain of *d*s and *g*s (Figure 16). What is the standard numeral for this chain?

Figure 16

Two important points about this last procedure are: (1) We have saved a step by dealing directly with *d* standard at the beginning of the exercise, and (2) we arranged the final chain with all the *d* strips together and all the *g* units together. Later in this chapter those two procedures will form the basis for doing the usual algorithms. An *algorithm* is a procedure for calculating based on the structure of the numeration system you are using.

You may do the exercises of this section using either of the two procedures outlined above: dealing with any of the strips or switching to *d* standard at the start of a problem.

EXAMPLE   Here is the procedure for switching to *d* standard for a *k* and *r* stack:

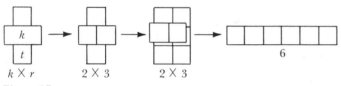

Figure 17

EXAMPLE   Here is an example using rose and white.

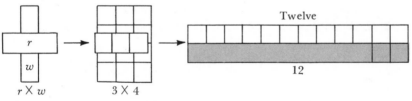

Figure 18

1. From the previous chapter you know that stacking and stretching out strips is analogous to multiplying whole numbers. Using the dark blue standard notation, find the products for Table 3 using this procedure:

a. Take two chains and make a stack.

b. Fill in the region under the top chain.

c. Find the dark blue standard chain for the strips of the region.

d. Record the result using dark blue standard notation.

Table 3   Multiplication in dark blue standard

×	1	2	3	4	5	6	7	8	9	10	11	12
1	1	2	3	4								
2	2	4										
3	3											
4												
5												
6												
7												
8												
9												
10												
11												
12												

2. Find the standard chain and express as a numeral.

EXAMPLE $2 \times 3 \times 5 = 30$

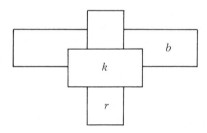

(a)

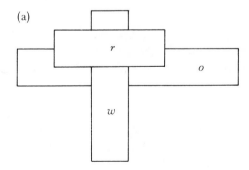

$3 \times 4 \times 6$

(b)

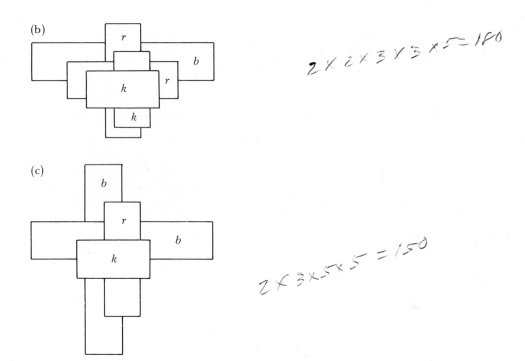

$2 \times 2 \times 3 \times 3 \times 5 = 180$

(c)

$2 \times 3 \times 5 \times 5 = 150$

3. Represent the numbers in the following exercises with strips, make a stack with them, find the dark blue standard chain, then write the resulting base-ten numeral.

EXAMPLE   2, 2, 5   <u> 20 </u>

   a. 3, 4, 5   <u> 60 </u>

   b. 4, 6, 6   <u> 144 </u>

   c. 2, 3, 5, 7   <u> 216 </u>

   d. 3, 3, 3, 3   <u> 81 </u>

   e. 5, 5, 3   <u> 75 </u>

   f. 5, 5, 5   <u> 125 </u>

   g. 2, 5, 2, 5   <u> 100 </u>

4. Check your answers for problem 3 with the answers to it on page 115. What is their relationship?

         same   numbers

5. Check the operation of multiplication on the set of whole numbers $\{0, 1, 2, 3, 4, 5, 6, \ldots\}$ for each of these properties:

a. Closure

yes

b. Associativity

yes

c. Existence of an identity element

yes

d. Existence of an inverse for each element

No

e. Commutativity

yes

5. Division

Make a chain that is represented by 20 (Figure 19).

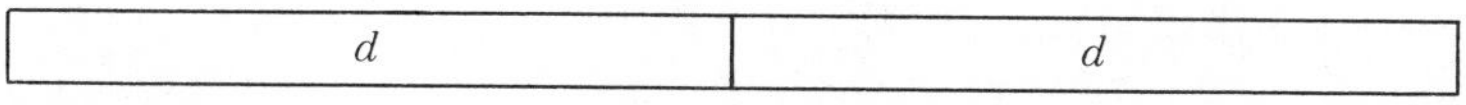

d	d

Figure 19

Our objective is to represent this chain with a *stack* of strips that includes a specific color. Here is an example with white and the 20 chain of Figure 19.

EXAMPLE Lay out white strips (Figure 20).

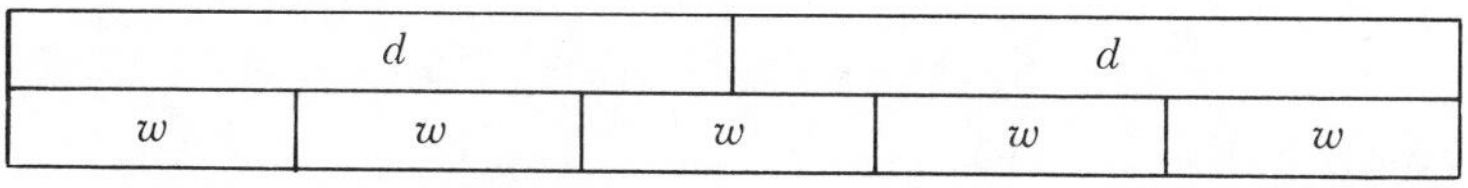

d			d	
w	w	w	w	w

Figure 20

Arrange the white strips of 20 so that they form a rectangular array that is light blue high (Figure 21).

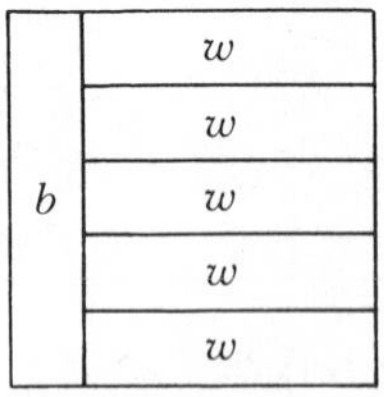

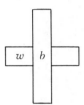

Figure 21

Remove all but one white and the light blue to make the stack (Figure 22). The result is a stack that includes a white strip. The "other" strip of the stack is light blue.

Figure 22

1. Make two-strip stacks with the specified color for each of the numbers given. (Sometimes one of the strips of a stack is represented by a chain of other strips, such as two dark blues and a green.)

EXAMPLE   Make a two-strip stack that includes a black for the number 8.

Given the chain in Figure 23, we must make an array of black strips as in Figure 24. Now make a stack as in Figure 25. The other strip of the stack is white.

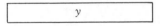

Figure 23

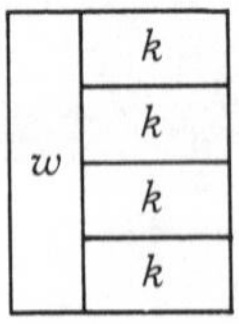

Figure 24

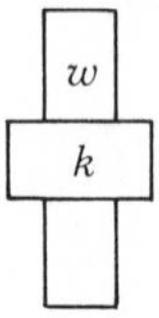

Figure 25

a. Make a two-strip stack that includes a rose strip for 12. The other strip is ___*white*___ .

b. Make a two-strip stack that includes an orange strip for 30. The other strip is ___*light blue*___

c. Make a two-strip stack that includes a light blue strip for 75. The other strip is ___*dark blue*___ *and a light blue*

d. Make a two-strip stack that includes a white strip for 31. The other strip is ___*impossible*___ .

2. Make stacks with as many nongreen strips as possible for these numbers:

EXAMPLE 1 ___16___ The most strips possible for this number is four blacks.

EXAMPLE 2 ___28___ The most strips possible for this number is three: two blacks and one purple.

EXAMPLE 3 ___21___ rose and purple

EXAMPLE 4 ___19___ *no stack possible*

EXAMPLE 5 ___30___ k, r, and b

a. $\dfrac{24}{\text{3 bl + 1 R}}$ b. $\dfrac{17}{\text{not poss}}$

c. $\dfrac{27}{\text{3 K}}$ d. $\dfrac{29}{\text{not poss}}$

e. $\dfrac{84}{\text{2 b's 1R, 1 pur}}$ f. $\dfrac{41}{\text{not poss}}$

g. $\dfrac{51}{\text{1 R (1 dk blk's pur)}}$ h. $\dfrac{79}{\text{not pos}}$

3. Complete Table 4 for the operation of finding the other strip of a stack. Select a number from the left column. Now represent this number with a two-strip stack that includes a number from the upper row. The other strip of the stack should be recorded in the table.

EXAMPLE The number 6 (from left column) can be represented by a stack that includes a black strip (2 from the top row). The other strip of the stack is a rose (3 recorded in the table).

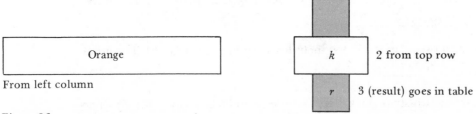

Orange					

From left column

	k	2 from top row
	r	3 (result) goes in table

Figure 26

Table 4   Find the other strip

Make a two stack using one of these numbers.

	1	2	3	4	5	6	7	8	9	10	11	12
1	1											
2	2	1										
3	3		1									
4	4	2		1								
5	5											
6	6	3	2									
7	7											
8	8	4		2								
9	9		3									
10	10	5			2							
11	11											
12	12	6	4	3		2						
13	13											
14		7										
15			5		3							
16		8		4								
17												
18		9	6			3			2			
19												
20		10		5	4					2		
21			7			3						
22		11									2	
23												
24		12	8	6		4		3				2
25					5							

Select one of these numbers.

Of course the selection of the *rose* strip from the top row would have given a result of black. Consequently, you should fill in the box across from 6 and down from 3 with a 2.

4. The following questions refer to Table 4, which you just completed.

  a. Some of the rows of Table 4 have *no* entries. Make a conjecture about these rows.

  ÷ by 1 and self

  b. Make a conjecture about the results from pairs of numbers you chose that were the same such as 2 and 2 or 8 and 8. What should the result be? Why?

  1   *n ÷ n = 1*

  c. Some of the rows of Table 4 have only *one* number entry. Make a conjecture about these rows.

  (prime)²

  d. Some of the rows of Table 4 have *several* number entries. Make a conjecture about these rows.

  ÷ by several Numbes,

  e. There are no entries in the upper right part of Table 4. Make a conjecture about this fact.

  no whole number which when multipled by a number in the Top row results in a number in the correspoding left colums,

  f. Make a conjecture about pairs of numbers such as 9 and 2 where it is impossible to use the 2 in a stack that represents 9. Devise a way to handle this situation to come up with a result.

  fractions

6. *Properties of numbers*

Each of the exercises in this section deals with an important concept of arithmetic. You should do the activities that develop these notions carefully, recording observations for use in making various conjectures.

1. Make a chain that represents 6. Can you make a chain of only black strips that is just as long? YES NO (Circle one.) Try to make chains of black strips that are as long as chains that represent the numbers in Table 5.

Table 5   Black chains

Number	Black chain as long
1	YES (NO)
2	(YES) NO
3	YES (NO)
4	(YES) NO
5	YES (NO)
6	(YES) NO
7	YES (NO)
8	(YES) NO
9	YES (NO)
10	(YES) NO
11	YES (NO)
12	(YES) NO
13	YES (NO)
14	(YES) NO
15	YES (NO)
16	(YES) NO
17	YES (NO)

a. Make a conjecture about the numbers in Table 5.

every other number

b. Devise a way to tell whether or not a given number can be represented with an all-black chain. Describe it with examples.

end in 0, 2, 4, 6, 8

c. Which of these numbers can be represented by an all-black chain?

(i)	19	YES	~~NO~~
(ii)	26	YES	NO
(iii)	277	YES	~~NO~~
(iv)	156	YES	NO
(v)	22,223	YES	~~NO~~
(vi)	333,332	YES	NO

d. Numbers that can be represented by an all-black chain are called *even*. Numbers that cannot be represented by an all-black chain are called *odd*. Devise a rule that identifies odd and even numbers represented with base-ten numerals.

0, 2, 4, 6, 8
1, 3, 5, 7, 9

2. Make a chain that represents 8 (Figure 27). Can you make a one-color chain that is just as long? (Do not use one single strip or just the green unit strips.)

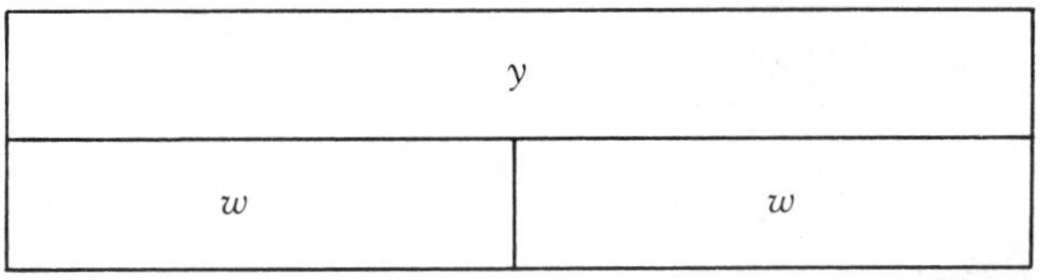

White

Figure 27

The answer to the question is *yes.* Figure 27 is a sketch of *one* possible solution. Name another solution. ___*black*___ (Color)

Find a one-color chain (not just units and more than one strip) for each of the following exercises:

a.

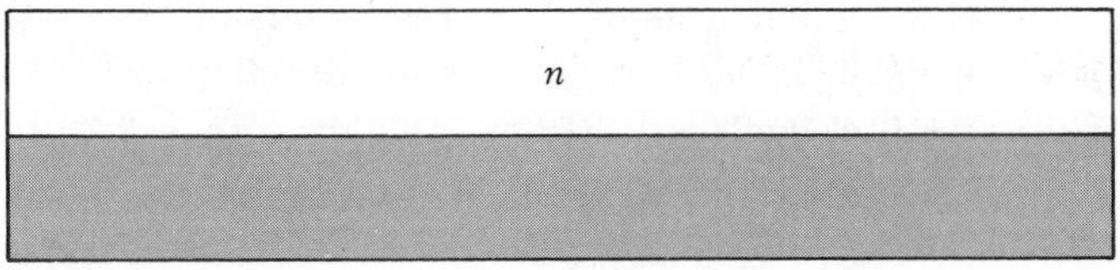

rose
Color

b.

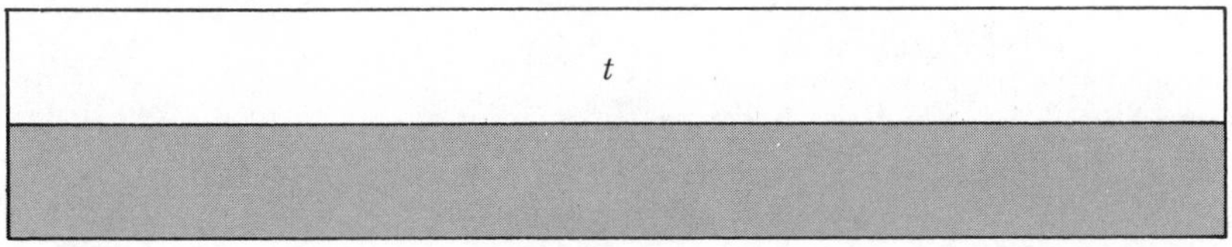

Impossible
Color

c.

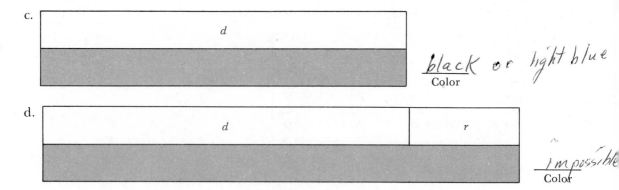

black or light blue
Color

d.

impossible
Color

e. You can find one-color chains for some numbers (such as 8 and 10) but not for others (such as 11 and 13). Circle all the numbers in Table 6 for which you *cannot* find one-color chains.

Table 6   One-color chains

①	⑪	21	㉛	㊶	51	61	㉛	81	91
②	12	22	32	42	52	62	72	82	92
③	⑬	㉓	33	㊸	53	63	㉝	㉘	93
4	14	24	34	44	54	64	74	84	94
⑤	15	25	35	45	55	65	75	85	95
6	16	26	36	46	56	66	76	86	96
⑦	⑰	27	㊲	㊼	57	67	77	87	㊾
8	18	28	38	48	58	68	78	88	98
9	⑲	㉙	39	49	㊾	69	㉛	89	99
10	20	30	40	50	60	70	80	90	100

The circled numbers (except the number 1) are called *primes*. By convention the number 1 is not considered a prime. The nonprimes are called *composites*.

f. A long time ago a famous mathematician made a conjecture about whole numbers and primes. It goes like this: Every nonprime number greater than one can be represented as a stack of primes. Demonstrate that this conjecture is true for these numbers:

i. 18   ii. 29
2×3×3

iii. 35   iv. 87
5×7   3×29

v. 155   vi. 182
5×31   2×7×13

3. Make two chains, one of black and one of light blue until they are the same length (Figure 28). Use as few strips as possible.

2	2	2	2	2
5		5		

Figure 28

Each of the chains in Figure 28 represents the number 10. This number is called the *least common multiple* (lcm) of 2 and 5. Find the least common multiple of these pairs of numbers:

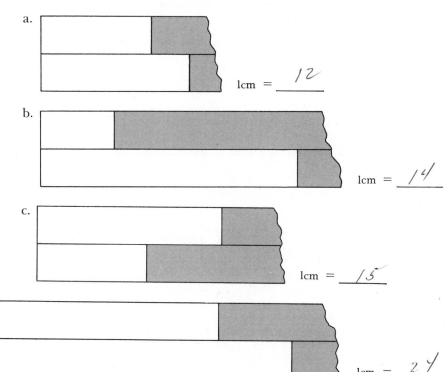

a.

lcm = _12_

b.

lcm = _14_

c.

lcm = _15_

d.

lcm = _24_

e. Find the least common multiple for these pairs of numbers:

i. 8, 12 lcm _24_   ii. 4, 8 lcm _8_

iii. 6, 9 lcm _18_   iv. 10, 15 lcm _30_

v. 10, 21 lcm _210_   vi. 25, 5 lcm _25_

4. Make chains for 12 and 15. Using the same color, cover the shaded regions below the two chains (Figure 29).

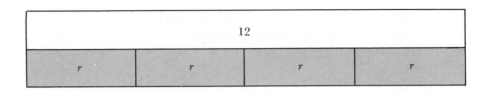

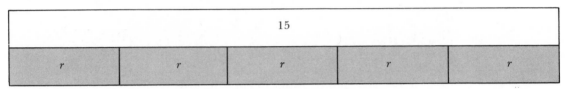

Figure 29

Rose is a color that can be used for both chains.

a. Use a similar procedure for the pair of chains that follows. Name the color used.

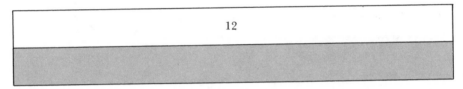

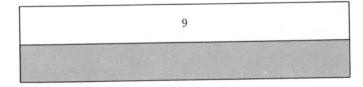

Name of color ___rose___

b. Find a common color for making chains as long as these:

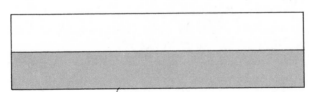

Name of color ___white___

You may have used black for each of these chains. You could have also used white. For this exercise, we use the white because they are longer than the black strip. The objective is to find the *greatest common color* (gcc) in making chains just as long as two or more numbers.

c. Find the greatest common color for making chains as long as these:

Name of color *light blue*

d. Find the greatest common color for making chains as long as these pairs of numbers:

    i. 24 and 16 gcc *yellow*

    ii. 20 and 16 gcc *white*

    iii. 18 and 12 gcc *orange*

    iv. 36 and 24 gcc *pink*

e. For larger numbers, the greatest common color designation is inappropriate. A *combination of strips* as a "common color" might be necessary. Consequently, a more general approach would be to identify the *common length* with a numeral. The name usually used is *greatest common factor* (gcf). Find the greatest common factor for these pairs of numbers.

    EXAMPLE 20 and 24 gcf   *4*

        i. 30 and 45   gcf *15*

        ii. 24 and 36   gcf *12*

        iii.   9 and 18   gcf *9*

        iv. 17 and 51   gcf *17*

5. Select a divisor from the top row of Table 7. Check to see if it divides into the numbers in the dividend column. Place a check (√) in the appropriate squares to indicate that the first number you chose was a *factor* of the second number.

Table 7   Quotients

Divisor

÷	2	3	4	5	6	7	8	9	10	11	12
4	✓		✓								
5				✓							
6	✓	✓			✓						
7						✓					
8	✓		✓				✓				
9		✓						✓			
10	✓								✓		
11										✓	
12	✓	✓	✓		✓						✓
13											
14	✓					✓					
15		✓		✓							
16	✓										
17											
18	✓										
19											
20	✓										
21						✓					
22	✓										
23											
24	✓										
25											
26	✓										

Dividend

EXAMPLE   Select the divisor 7. Since 7 is certainly not a factor of 4, 5, or 6, there will be no checks in the first three boxes. But 7 *is* a factor (along with 1) of 7 so that you should place a check in that box. The next check would appear across from 14 and another across from 21.

a. Complete the table for all the other divisors. The 2 column has been done for you.

b. Devise a system for predicting when a number, expressed as a base-ten numeral, is divisible by 2.

even

c. Devise a system for predicting when a number, expressed as a base-ten numeral, is divisible by 3.

sum ÷ by 3

d. Devise a system for predicting divisibility of numbers, expressed in the base ten, for:

   i. 4   *last 2 digits ÷ 4*

   ii. 5   *0, 5*

   iii. 6   *2 & 3*

   iv. 7   *easier*

   v. 8   *Last 3 digits ÷ 8*

   vi. 9   *sum digits ÷ 9*

vii. 10   *end in 0*

viii. 11   *add alternate digits*
                 subt

ix. 12   *3, 4*

x. 100   *end 00*

Summary

At this point you should be able to understand what you are accomplishing with this text. The mathematics should clearly stand out through modeling with the strips. In fact, you should begin thinking that everything mathematical in your life could be modeled by some manipulative device or a real situation. Mathematics isn't magic. It is based on the world around you and can be modeled by various devices.

Although you will not be expected to remember (for long) the procedures for modeling greatest common factor, you should remember that it *can* be modeled and it *is* a reasonable and simple idea. This is a very important point because it will give you faith and confidence in mathematics. In the end, mathematics should be your friend.

References

Bidwell, James. "Learning Structures for Arithmetic." *The Arithmetic Teacher,* 16 (April 1969), 263–268.

Cacha, Frances B. "Understanding Multiplication and Division of Multidigit Numbers." *The Arithmetic Teacher,* 19 (May 1972), 349–354.

D'Augustine, Charles. "Multiple Methods of Teaching Operations." *The Arithmetic Teacher,* 16 (April 1969), 259–262.

Grinther, John L. "Some Activities with Operation Tables." *The Arithmetic Teacher,* 15 (December 1968), 715–717.

Hervey, Margaret A., and Bonnie H. Litwiller. "A Graphical Representation of Multiples of Whole Numbers." *The Arithmetic Teacher,* 18 (January 1971), 47–48.

Kennedy, Leonard M. "Organizing Composite and Prime Numbers." *The Arithmetic Teacher,* 11 (February 1964), 109–111.

Lichtenberg, Betty Plunkett. "Zero Is an Even Number." *The Arithmetic Teacher,* 19 (November 1972), 535–538.

8

Integers

There are many instances where integers are useful in the real world. If you are planning a picnic and the weather report predicts a change in temperature of 20°, you want to know the direction of that change. A force of 100 lb would be helpful in lifting a safe only if the force were in the same direction in which you were lifting. Winning a 2000 mi trip is useful if it's in a direction you wish to go. A balance of $200 in your checking account is much more pleasing than a $200 overdraft.

Although not many people use formal operations with signed numbers in their daily lives, they do use the concepts. It is therefore useful for us to develop a concrete basis for understanding and remembering just how integers and the operations of addition, subtraction, multiplication, and division of integers work.

1. Left and right trips

Label a row of pegs on the geoboard as in Figure 1.

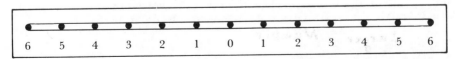

6   5   4   3   2   1   0   1   2   3   4   5   6

Figure 1

Starting at zero, stretch a rubber band 3 units to the *right* as in Figure 2. From that peg, stretch another rubber band 5 units to the *left*. How many units from zero and on what side is the peg to which the second rubber band was stretched?

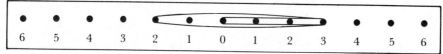

Figure 2

We can simplify the instructions above to $3R + 5L$, where $+$ means "stretch" in the direction indicated. The result is $2L$, so the entire relation is $3R + 5L = 2L$.

1. Add the following on the geoboard and record the results as in Figure 3.

a. $2L + 4L =$ __*6 L*__   b. $4R + 0R =$ __*4 R*__

c. $3R + 2R =$ __*5 R*__   d. $5R + 1R =$ __*6 R*__

e. $3L + 3L =$ __*6 L*__   f. $4L + 1L =$ __*5 L*__

Example: $5R + 4L = 1R$

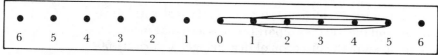

Figure 3

2. Make a conjecture about the results of adding lefts to lefts or of adding rights to rights. Make up your own examples to check your conjecture.

$$L + L = L$$
$$R + R = R$$

3. Now mix the directions in which the rubber bands are stretched. Add the following on the geoboard and record.

a. $2L + 4R =$ __*2 R*__   b. $5R + 3L =$ __*2 R*__

c. $5L + 0R =$ __*5 L*__   d. $3R + 3L =$ __*0*__

e. $5L + 2R =$ __*3 L*__   f. $3R + 6L =$ __*3 L*__

4. Make a conjecture about the results (units and directions) of the operations in problem 3. Make up your own examples to check your conjecture.

Larger   Number

5. We can replace R and L with signs for numbers. Numbers to the left of zero (less than zero) are called *negative* numbers, and numbers to the right of zero (greater than zero) are called *positive* numbers. Instead of writing $2R$, we write $^+2$, instead of writing $3L$ we write $^-3$. Remember that $^+2$

represents both quantity and direction, whereas 2 represents quantity only. These are *signed numbers* and we say "positive two" for $^+2$ and "negative three" for $^-3$.

Place a + in front of the numbers used to label the pegs on the right side and a − in front of those on the left, as in Figure 4. Note that 0 is neither positive nor negative. Note also how the following signed numbers are added on the geoboard.

a. $^-3 + ^-2 = $ __−5__   b. $^-6 + ^+4 = $ __− 2__

c. $^-3 + ^+3 = $ __0__   d. $0 + ^-4 = $ __− 4__

e. $^+4 + \ 0 = $ __4__   f. $^+5 + ^-2 = $ __3__

g. $^+3 + ^+2 = $ __5__   h. $^-3 + ^+5 = $ __2__

Example: $^+4 + ^-5 = ^-1$

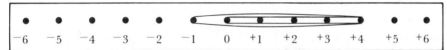

| $^-6$ | $^-5$ | $^-4$ | $^-3$ | $^-2$ | $^-1$ | 0 | $^+1$ | $^+2$ | $^+3$ | $^+4$ | $^+5$ | $^+6$ |

Figure 4

6. Complete Table 1 without the geoboard. Then check your results on the geoboard. Note that the table is only a sample of the many possible signed numbers that could be included.

Table 1 Adding integers

+	$^-3$	$^-2$	$^-1$	0	$^+1$	$^+2$	$^+3$
$^-3$	−6	−5	−4	−3	−2	−1	0
$^-2$	−5	−4	−3	−2	−1	0	1
$^-1$	−4	−3	−2	−1	0	1	2
0	−3	−2	−1	0	1	2	3
$^+1$	−2	−1	0	1	2	3	4
$^+2$	−1	0	1	2	3	4	5
$^+3$	0	1	2	3	4	5	6

7. Explore the following properties for the operation of addition on the set of signed numbers. Give examples of each.

a. Identity element   *0*

b. Inverses $3, {}^-3$

c. Associative property

$$(^-3 + 1) + 2 = 0 \qquad ^-3 + (1 + 2) = 0$$

d. Commutative property

$$(^-2 + 3) = (3 + {}^-2)$$

8. The set of numbers $\{\ldots, {}^-4, {}^-3, {}^-2, {}^-1, 0, {}^+1, {}^+2, {}^+3, {}^+4, \ldots\}$ goes on indefinitely in both directions. We call these numbers *integers*. Imagine a row of pegs on the geoboard continuing indefinitely in both directions from zero. Based on your conjectures about adding signed numbers, find these sums:

a. $^+23 + {}^-16 = \underline{7}$ b. $^-31 + {}^-23 = \underline{-54}$

c. $^+34 + {}^-34 = \underline{0}$ d. $^+27 + {}^-19 = \underline{8}$

e. $^+8 + {}^-11 = \underline{-3}$ f. $^+18 + {}^+14 = \underline{32}$

g. $^-8 + {}^-7 = \underline{-15}$ h. $^-16 + {}^+9 = \underline{-7}$

9. Determine if the set of integers and the $+$ operation form a group. Explain.

yes

2. Reversing direction

Starting at zero, stretch a rubber band $^+6$ units (that is, 6 units to the right); from that peg, stretch another rubber band the opposite of $^+3$ (that is, 3 units to the left). The result of this is given in Figure 5.

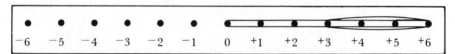

Figure 5

1. If you start with $^+5$ and stretch another rubber band the opposite of $^+7$, you end up at $^-2$. Try these combinations and indicate the end point.

a. $^+5$ opposite $^+2$ ___**3**___   b. $^+8$ opposite $^+7$ ___**1**___

c. $^+4$ opposite $^+4$ ___**0**___   d. $^-3$ opposite $^+6$ ___**-9**___

e. $^-9$ opposite $^-5$ ___**-4**___   f. $^+6$ opposite $^-1$ ___**7**___

g. $^-2$ opposite $^+6$ ___**-8**___   h. $^-7$ opposite $^-7$ ___**0**___

2. As you may have suspected, the opposite operation is analogous to subtraction. We shall name it "subtraction" and use the $-$ symbol. Consequently, we shall have $^+5 - ^+2 = ^+3$ in place of $^+5$ opposite $^+2$ _____. Carefully complete these pairs of problems on your geoboard.

a. $^+3 - ^+2 =$ ___**1**___   b. $^-2 - ^+4 =$ ___**-6**___

$^+3 + ^-2 =$ ___**1**___ $^-2 + ^-4 =$ ___**-6**___

c. $^+6 - ^+4 =$ ___**2**___   d. $^+2 - ^+5 =$ ___**-3**___

$^+6 + ^-4 =$ ___**2**___ $^+2 + ^-5 =$ ___**-3**___

e. $^-4 + ^-3 =$ ___**-7**___   f. $^+5 - ^+3 =$ ___**2**___

$^-4 - ^+3 =$ ___**-7**___ $^+5 + ^-3 =$ ___**2**___

g. $^+3 - ^-2 =$ ___**5**___   h. $^-6 + ^+4 =$ ___**-2**___

$^+3 + ^+2 =$ ___**5**___ $^-6 - ^-4 =$ ___**-2**___

3. Subtract the following by making appropriate sketches on the geoboard graphs as shown in Figure 6.

Example: $^+3 - ^-2 = ^+5$, we add the inverse of $^-2$ by stretching the opposite direction of $^-2$; that is, $^+3 + ^+2 = ^+5$

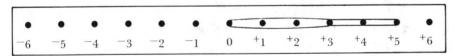

Figure 6

a. $^+2 - ^-4 =$ ___**6**___

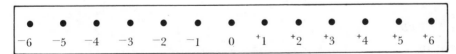

b. $^-5 - ^-3 =$ ___**-2**___

c. $0 - ^-5 =$ ___**5**___

d. $^+4 - {}^-1 = $ __5__

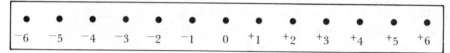

e. $^-3 - {}^-5 = $ __2__

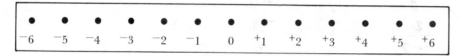

f. $^+1 - {}^-2 = $ __3__

4. Make a conjecture about subtracting negative numbers. Make up some of your own problems to check your conjecture.

$$a - {}^-b = a + b$$

5. Imagine again that a row of pegs on the geoboard continues indefinitely in both directions from zero. Based on your conjectures about subtracting integers, perform the following operations.

a. $^+23 - {}^+32 = $ __−9__   b. $^-13 - {}^-9 = $ __−4__

c. $^+8 - {}^-28 = $ __36__   d. $^-46 - {}^+10 = $ __−56__

e. $^+146 - {}^+192 = $ __−46__   f. $^-31 - {}^-49 = $ __18__

g. $^+17 - {}^+17 = $ __0__   h. $^-103 - 0 = $ __−103__

3. A jumping series

Starting at zero, make a series of jumps of two units to the *right*, and record in Figure 7. A record of the landing points is given in Table 2.

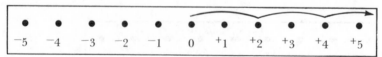

Figure 7

1. Continue the pattern for this procedure. Note that the size of the jump is $^+2$ since it is two units to the right. The number of jumps is positive because we assume a forward jump. We indicate a forward jump as a positive integer.

Table 2   Positive jumps

Number of jumps	Size of jumps	Landing point
0	⁺2	0
⁺1	⁺2	⁺2
⁺2	⁺2	⁺4
⁺3	⁺2	⁺6
⁺4	⁺2	8
⁺5	⁺2	10
⁺6	⁺2	12

2. In Figure 8 start at zero and make a series of jumps of three units to the *left*. Record the results in Table 3 and continue the pattern.

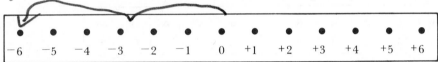

Figure 8

Table 3   Negative jumps

Number of jumps	Size of jumps	Landing point
0	⁻3	0
⁺1	⁻3	⁻3
⁺2	⁻3	-6
⁺3	⁻3	-9
⁺4	⁻3	-12
⁺5	⁻3	-15
⁺6	⁻3	-18

3. Continue the pattern of Table 4, in which the first column decreases in size. Describe a physical action for a negative number of jumps.

4. The action of forward or backward jumps of a specific size and direction models a multiplicative operation. Perhaps you have already guessed that 3 forward jumps of ⁻3 distance and direction can be described as ⁺3 × ⁻3 = ⁻9. Do each of the following problems on your geoboard and record the jumps in the space provided.

Table 4   Backward jumps

Number	Size	Landing
+2	+4	+8
+1	+4	+4
0	+4	0
−1	+4	−4
−2	+4	−8
−3	+4	−12
−4	+4	−16
−5	+4	−20

a. $^+2 \times {}^+3 =$ __6__

b. $^+2 \times {}^-3 =$ __−6__

c. $^-2 \times {}^+3 =$ __−6__

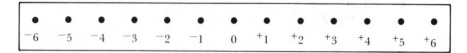

d. $^-2 \times {}^-3 =$ __6__

5. Continue the pattern in Table 5. Make a conjecture about multiplying integers. Give examples to justify your conjecture. Now complete Table 6 using your conjecture.

6. Calculate the following products. Use your geoboard if you are unsure of your products.

Table 5   Negative two

$^+3$	X	$^-2$	=	$^-6$
$^+2$	X	$^-2$	=	-4
$^+1$	X	$^-2$	=	-2
0	X	$^-2$	=	0
$^-1$	X	$^-2$	=	2
$^-2$	X	$^-2$	=	4
$^-3$	X	$^-2$	=	6
$^-4$	X	$^-2$	=	8

Table 6   Negative four

$^+2$	X	$^-4$	=	$^-8$
$^+1$	X	$^-4$	=	-4
0	X	$^-4$	=	0
$^-1$	X	$^-4$	=	4
$^-2$	X	$^-4$	=	8
$^-3$	X	$^-4$	=	12
$^-4$	X	$^-4$	=	16
$^-5$	X	$^-4$	=	20

a. $^+2 \times {}^+3 =$ __6__

b. $^+2 \times {}^-3 =$ __-6__

c. $^-2 \times {}^+3 =$ __-6__

d. $^-2 \times {}^-3 =$ __6__

e. $^-6 \times {}^-5 =$ __30__

f. $^+7 \times {}^-2 =$ __-14__

g. $^-2 \times {}^+7 =$ __-14__

h. $^+2 \times {}^-7 =$ __-14__

i. $^+2 \times {}^-3 \times {}^-5 =$ __30__

j. $^-2 \times {}^-3 \times {}^+5 =$ __30__

k. $^-2 \times {}^+3 \times {}^-5 =$ __30__

l. $^-2 \times {}^-3 \times {}^-5 =$ __-30__

7. Continue filling out Table 7 for products of the indicated integers.

8. Explore each of these properties for multiplying integers. Give at least six examples before making a conjecture.

a. Closure   *yes*

Table 7   Integer products

X	⁻10	⁻9	⁻8	⁻7	⁻6	⁻5	⁻4	⁻3	⁻2	⁻1	0	⁺1	⁺2	⁺3	⁺4	⁺5	⁺6	⁺7	⁺8	⁺9	⁺10
⁺10	-100																				
⁺9																					
⁺8																					
⁺7																					
⁺6																					
⁺5																					
⁺4																					
⁺3														⁺9							
⁺2											0										
⁺1			⁻8										⁺2								
0								0													
⁻1																					
⁻2								⁺6					⁻4								
⁻3																					
⁻4																					
⁻5																					
⁻6																					
⁻7												⁻7									
⁻8																					
⁻9																					
⁻10																					

b. Identity element

1

c. Inverses

no   only   1 & -1

d. Associative property

yes

e. Commutative property

yes

9. Determine whether the set of integers and the operation of multiplication form a group. Explain.

No   each element does not have inverse

10. Determine whether the set of integers and the operation of multiplication form a field. Use the following questions as a guide. *No*

a. Does addition on the integers form a group?

yes

b. Does multiplication on the integers (without zero) form a group?

No

c. Does the distributive property hold for multiplication over addition on the integers?

$a \times (b + c) = (a \times b) + (a \times c)$ *yes*

4. *Finding the lengths of jumps*

An interesting puzzle can be generated by giving only the finish point and the number of jumps, with the length of the jump to be determined. Complete Table 8 before making a rule for determining the sign of the length of the jump.

1. Use the rule you just generated to find the length of the jumps in Table 9.

2. The last four entries in Table 9 could not be completed because our jumps (up until now, at least) have been from peg to peg. You cannot get to $^{+}15$ in $^{+}4$ jumps, because jumps of $^{+}3$ are too short and jumps of $^{+}4$ are too long. Can you find a finish peg that cannot be reached with *more than one jump* and *jumps of $^{+}2$ units or more*? If you assume a very long line of pegs, there are many of these pegs.

Table 8   Length puzzle

Finish	Number of jumps	Length of jumps
⁻8	⁺4	-2
⁻8	⁻4	2
⁺9	⁻3	-3
⁻10	⁺2	-5
⁻12	⁻3	4
⁺18	⁻9	-2
⁺28	⁺7	4
⁻32	⁺8	-4
⁺48	⁻12	4

Table 9   Signs pattern

Finish	Number of jumps	Length of jumps
⁺6	⁺2	3
⁺6	⁻2	-3
⁻6	⁺2	-3
⁻6	⁻2	3
⁺12	⁺6	2
⁺12	⁻6	-2
⁻12	⁺6	-2
⁻12	⁻6	2
⁺15	⁺4	
⁺15	⁻4	
⁻15	⁺4	
⁻15	⁻4	

3. One way we commonly express the relationship of the numbers in Tables 7 and 8 is to use the division sign. If you finish at ⁻12 after ⁻3 jumps, you had jumps of ⁺4. This is written ⁻12 ÷ ⁻3 = ⁺4. Find the size of the jumps for these expressions:

a. $^+9 \div {}^+3 =$ __**3**__　　　b. $^-10 \div {}^-2 =$ __**5**__

c. $^-12 \div {}^+4 =$ __**- 3**__　　d. $^+14 \div {}^-7 =$ __**- 2**__

e. $^-18 \div {}^+2 =$ __**-9**__　　f. $^-20 \div {}^-5 =$ __**4**__

g. $^+36 \div {}^+12 =$ __**3**__　　h. $^+120 \div {}^-30 =$ __**- 4**__

4. An interesting problem to consider involves the meaning of the expression:

$^+12 \div 0 =$ __*imp*__ .

What could be the size of the jumps for this relation? How should this problem be handled?

5. An important idea in mathematics is existence and uniqueness for operations. That is, if we use an operation (such as addition) on two numbers (such as $^+4$ and $^+7$), the result must exist and it must be unique. It would be most unfortunate to have the result of operations on two objects vary or not exist! Consequently, we can reject the idea of dividing by zero. Either it does not give a unique result (in the case of $0 \div 0$), or it does not exist (as in the case of $^+12 \div 0$). In fact, the rule is *never divide by zero.*

6. Check these properties for dividing integers. Remember to exclude division by zero.

   a. Closure

   No

   b. Associative property

   No

   c. Identity element

   yes!

d. Inverses

yes   except 0

e. Commutative property

No

Summary

Many ways exist to generate a meaningful foundation for understanding positive and negative numbers. Likewise, the operations of addition, subtraction, multiplication, and division can be modeled by a variety of situations. We have chosen a number-line approach that models trips in opposite directions because it is meaningful and easy to generate on a geoboard.

You should explore the other approaches described in the references for this chapter to see the variety of concrete situations in which positive and negative numbers are easily found. In each case the development of the rules for the operations must be logical and obvious; otherwise the model isn't much good.

Now you can ask yourself just how well you understand the rules for operations on integers from using the models in this chapter. You see, the objective wasn't just to get the answers to the exercises. Instead, we wanted you to be able to use integers in an accurate and meaningful way and to gain understanding that will be useful in all areas of number systems and their uses.

As you work in the next chapter, try to see the practicality of using two different models (colored strips and a geoboard) to generate fractions. First we shall develop the colored strip idea, then move to the geoboard. In both cases we have a model that accurately describes fractions and operations on fractions.

References

Ashlock, Robert, and Tommie West. "Physical Representation for Signed-Number Operations." *The Arithmetic Teacher,* 14 (November 1967), 549–554.

Cohen, Louis. "A Rationale in Working with Signed Numbers." *The Arithmetic Teacher,* 12 (November 1965), 563–567.

Hill, Warren H., Jr. "A Physical Model for Teaching Multiplication of Integers." *The Arithmetic Teacher,* 15 (October 1968), 525–528.

Hollis, Loge. "Multiplication of Integers." *The Arithmetic Teacher,* 14 (November 1967), 555–556.

Mauthe, Albert. "Climb the Ladder." *The Arithmetic Teacher,* 16 (May 1960), 354–356.

9

Rational numbers

The numbers that most people use most frequently are *rational numbers*. These numbers usually express the quantitative value of some object or event, or they express a relationship. The wide range of meaning and complexity of the numbers themselves requires a certain amount of sophistication on the part of users. To develop an understanding and appreciation of this number system, we will use several different approaches in this chapter.

The first approach will be based on relationships between pairs of colored centimeter strips. These relationships will be represented with a dark blue standard (base-ten) system through a division procedure. This procedure generates a decimal representation system that can be modeled with the flats, strips, and units of dark blue standard.

Of course there are other systems of naming rational numbers. Studying such systems emphasizes the clever nature of a base and place value system of naming fractional numbers. Consequently, we shall delve into base-five representations of fractions that are similar to our decimal system.

Then we shall switch to a geoboard model for rational numbers. This geoboard model uses the ratio of horizontal change to vertical change to represent a fraction. Normally this slope is identified with the rational number, but we shall use it to represent fractions.

All of this should help develop a thorough and meaningful understanding of the rational number system.

1. *White ones*

In previous chapters we have used the colored strips to represent whole numbers by letting the green square be one unit. Actually we can permit any strip to be a unit as long as we have specified it clearly. In this section

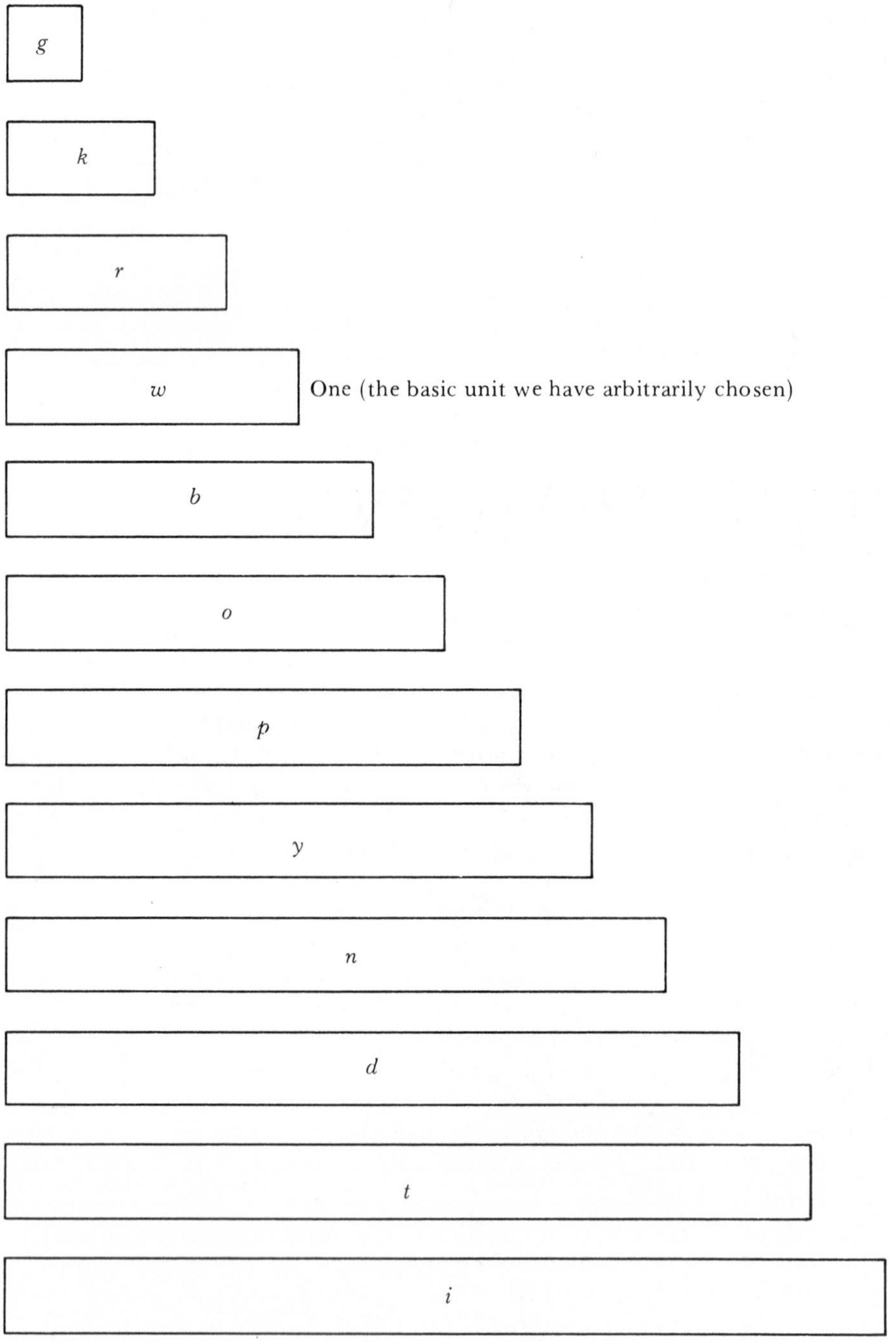

Figure 1

we shall use the white strip as our basic unit. All our numbers will be based on this white unit. We assume that you are now at an abstract level of operation with respect to the base-ten numeration system and can write numerals such as 12 and 23 without thinking of dark blue strips and green units. This is important because in this section you will be asked to use a dark blue strip as a representative of $2\frac{1}{2}$ while at the same time be writing the numeral 10.

1. If white is our unit and represents the number 1, name the set of 12 strips in Figure 1.

2. Since you already know several types of names for fractions, you may have named the strips differently from some of your classmates. To communicate between ourselves, we shall decide on a standard way of naming these numbers. First, if the numbers are whole (that is, white, yellow, and pink for 1, 2, and 3, respectively), there is no problem. But if they are not whole, we shall use this system of naming: *Relate the strip in question to the standard white unit.* Consequently, black would be one-half ($\frac{1}{2}$) and light blue would be one and one-fourth ($1\frac{1}{4}$) or five-fourths ($\frac{5}{4}$). Fill out Table 1 with the appropriate names.

Table 1 White one names

Color	Verbal name	Written numeral
g		$\frac{1}{4}$
k	one-half	$\frac{1}{2}$
r		$\frac{3}{4}$
w	one	1
b	five-fourths	$\frac{5}{4}$
o		$\frac{6}{4} = \frac{3}{2}$
p		$\frac{7}{4}$
y		$\frac{8}{4} = 2$
n		$\frac{9}{4}$
d		$\frac{10}{4} = 2\frac{1}{2}$
t		$\frac{11}{4}$
i		$\frac{12}{4} = 3$

3. What is the numerical value of the rose strip? The brown strip? Now find the numerical value of a rose and brown chain (Figure 2).

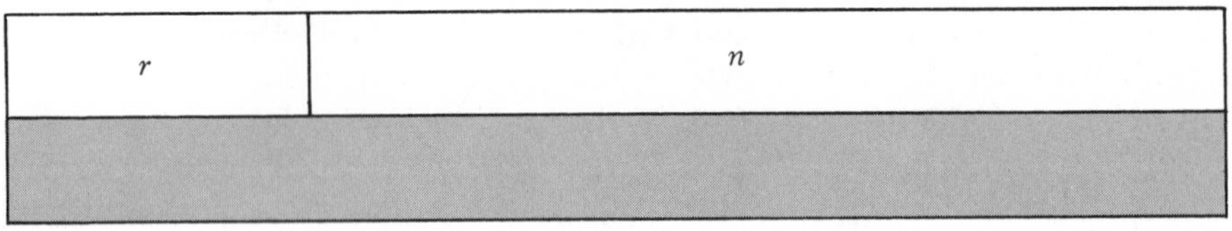

Figure 2

Find the numerical value of these chains:

(a)

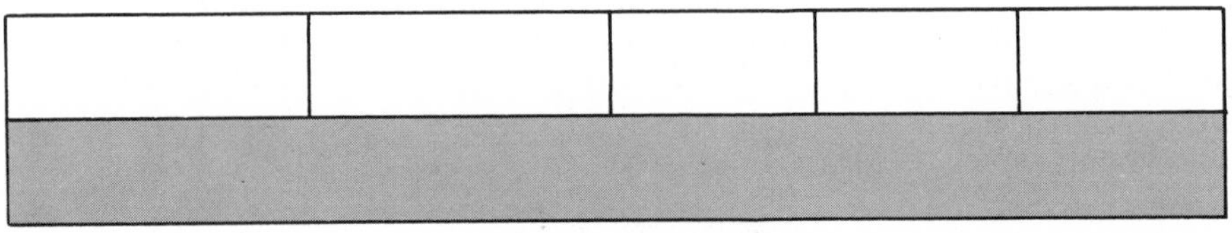

3

(b)

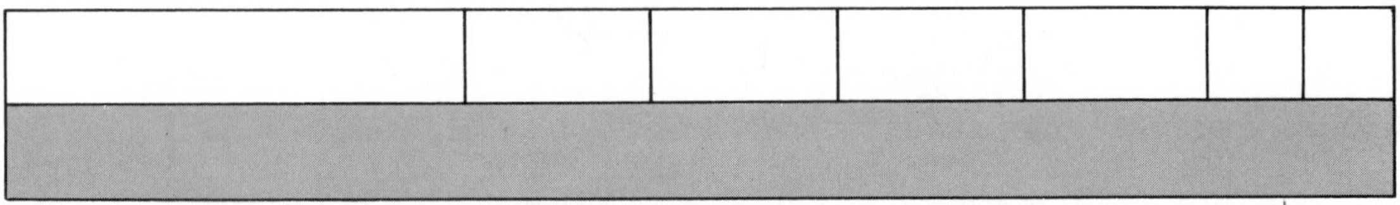

3¾

(c)

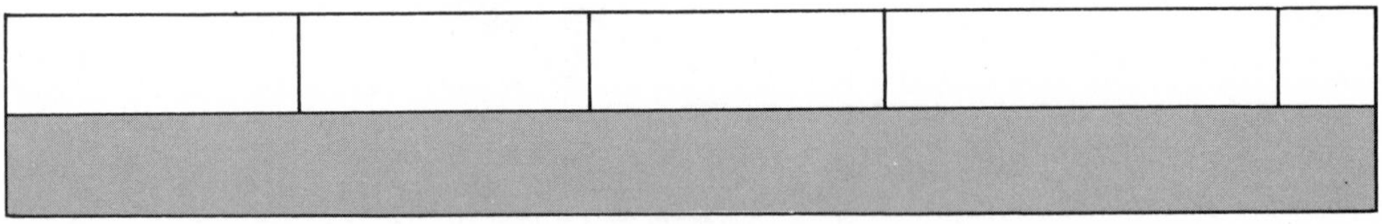

3½

(d)

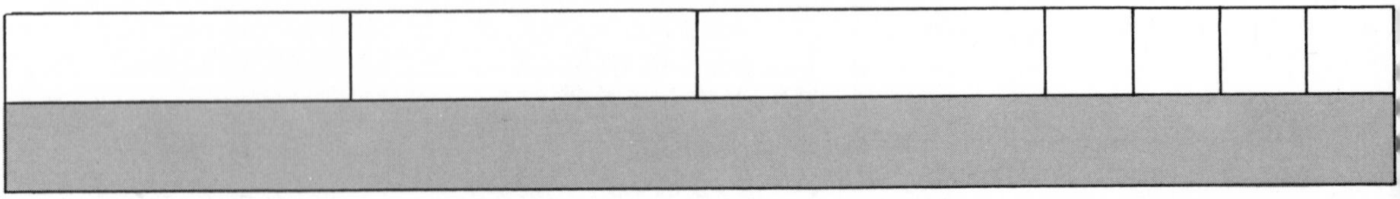

4

(e)

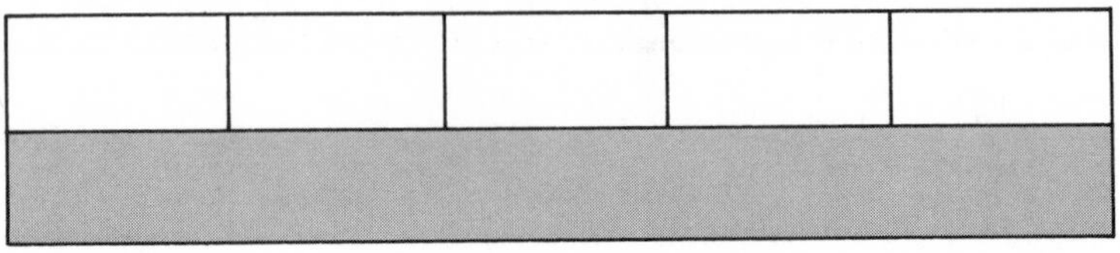

9½

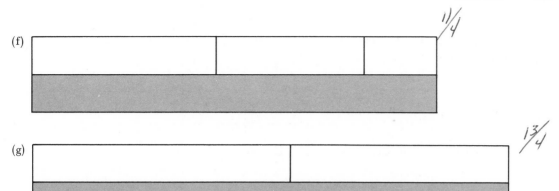

(f)   1¼

(g)   1¾

(h)   3

4. Fill in Table 2 with all possible results of adding 2 of the 12 colored strips. Record the numerals in the table.

EXAMPLES $w + w = 1 + 1 = 2$

$b + o = \frac{5}{4} + \frac{6}{4} = \frac{11}{4}$

$n + y = 2\frac{1}{4} + 2 = 4\frac{1}{4}$

5. How would you represent these numbers using the white strips as the unit? What restrictions does this system have?

  a. 6 $i + i$

  b. 7½ $i + i + o$

  c. 20¼ $i + i + i + i + i + i + n$

  d. 13¾ $i + i \; i + i + p$

  e. ⅓ imp
                imp

  f. 5⅕

6. Using Table 2 as a basis, answer these questions about adding white-one fractions:

  a. Is there closure for this system?   yes

  b. Is addition associative for these numbers?   yes

Table 2 Addition of white one fractions

+	g	k	r	w	b	o	p	y	n	d	t	i
g	$\frac{1}{2}$	$\frac{3}{4}$	1	$\frac{5}{4}$	$\frac{3}{2}$							
k												
r												
w				2								
b						$\frac{11}{4}$						
o												
p												
y												
n								$4\frac{1}{4}$				
d												
t												
i												

c. Is there an identity element for addition? *no*

d. Are there additive inverses? *no – no identity element*

e. Is addition commutative for these numbers? *yes*

7. Using a procedure similar to subtraction of whole numbers, describe a sequence of steps for using the strips as a model for these problems. Trace around your arrangements of strips and label them clearly.

a. $3 - 2 = 1$

b. $2\frac{3}{4} - 1\frac{1}{4} = 1\frac{2}{4}$

c. $4\frac{1}{2} - 1\frac{3}{4} = 2\frac{3}{4}$

d. $12\frac{1}{4} - 6\frac{2}{4} = 5\frac{3}{4}$

8. Make some conjectures about using the strips for subtraction where the white strip is the unit. *need To change units ofTen.*

9. Figure 3 represents a product. If our unit is white, this product would be $2\frac{1}{2}$ times 3. Use your strips to calculate the result of this multiplication. Is the product correct?

25, No

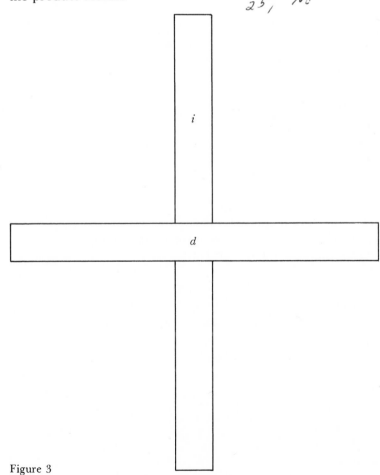

Figure 3

10. There is a fundamental flaw in the reasoning that would assign the previously used "system" of stacking strips to our "white-ones" numbers. It is this: In our whole number system, the unit was the green strip. Consequently, *every* other strip had one dimension that was 1 unit. With the white strip as the unit, the strips have one dimension that is $\frac{1}{4}$ unit. Thus the stacking procedure will not work.

Think of a way to modify the stacking procedure so that it will work as a means of representing multiplication with the white-ones system.

each strip needs To be "white" wide, i e 4 white units in width

2. Many ones

Although "white-ones" is a nice way to name fractions with denominators of four, it isn't useful for many other fractions. Using the light blue strip as a unit would give us fractions with denominators of 5. The dark blue as the unit gives denominators of 10. The entire set of strips would give fractions with denominators of 2, 3, 4, 5, 6, 7, 8, 9, 10, 11, and 12. Combinations of strips would give larger denominators.

1. If yellow is one, name the other strips.

$g = \underline{1/8}$ $b = \underline{5/8}$ $n = \underline{9/8}$

$k = \underline{1/4}$ $o = \underline{3/4}$ $d = \underline{5/4}$

$r = \underline{3/8}$ $p = \underline{7/8}$ $t = \underline{11/8}$

$w = \underline{1/2}$ $y = \underline{1}$ $i = \underline{3/2}$

2. If pink is the one, name the other strips.

$g = \underline{1/4 \; 2}$ $b = \underline{5/12}$ $n = \underline{3/4}$

$k = \underline{1/6}$ $o = \underline{1/2}$ $d = \underline{5/6}$

$r = \underline{1/4}$ $p = \underline{7/12}$ $t = \underline{11/12}$

$w = \underline{1/3}$ $y = \underline{2/3}$ $i = \underline{1}$

3. To keep fractional numbers from getting mixed up, use this procedure: Each time you represent a fractional number, use *both* the unit strip *and* the strip for the fractional number. Arrange them as in Figure 4.

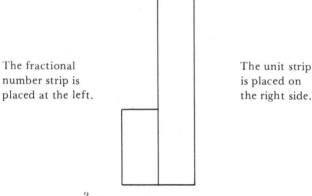

The fractional number strip is placed at the left.

The unit strip is placed on the right side.

Figure 4 $\frac{2}{5}$ The ratio is two to five.

Name these fractional numbers.

(a) 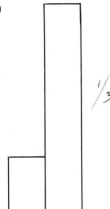 $\frac{1}{3}$

(b) 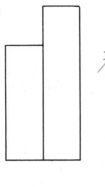 $\frac{3}{4}$

(c)  $\frac{4}{5}$

(d)  $\frac{5}{3}$

4. Represent these fractional numbers by naming the strips in the order they appear, left to right.

EXAMPLE $\frac{1}{4}$; ___*g*___, ___*w*___

    a. $\frac{2}{5}$ ___*K*___, ___*b*___

    b. $\frac{6}{2}$ ___*o*___, ___*K*___

    c. 2 ___*K*___, ___*g*___

    d. $\frac{3}{7}$ ___*r*___, ___*p*___

    e. $\frac{9}{10}$ ___*u*___, ___*d*___

5. Represent each of these fractional numbers with the strips. Keep each pair until all six examples are complete.

    a. $\frac{1}{2}$ b. $\frac{2}{4}$ g/K w/y

    c. $\frac{3}{6}$ d. $\frac{4}{8}$ K/w b/d

    e. $\frac{5}{10}$ f. $\frac{6}{12}$ r/o o/i

6. Make a conjecture about the relationship between the six pairs of strips in problem 5. *equal*

7. Find four pairs of strips that represent the relationship of one to three (1/3).

a. ___*g*___ , ___*r*___

b. ___*k*___ , ___*0*___

c. ___*r*___ , ___*n*___

d. ___*w*___ , ___*i*___

8. Describe another collection of strip pairs where the pairs all are in the same ratio. *g/w   k/y   r/i*

9. The numbers $\frac{1}{4}, \frac{2}{8}, \frac{3}{12}, \frac{4}{16}$, etc., are called equivalent fractions. Represent each of these fractions with the strips and compare the lengths of the two strips of each pair. Make a conjecture about your representations of these numbers.

$$\frac{1}{4} = \frac{2}{8} \qquad \frac{1}{4} \times \frac{2}{2} = \frac{2}{8}$$

10. Name some equivalent fractions for the following. (The appendix contains the game "Colored Squares," which is based on recognizing equivalent fractions.)

a. $\frac{1}{2}$ b. $\frac{2}{3}$

c. $\frac{1}{3}$ d. $\frac{3}{4}$

e. $\frac{1}{4}$ f. $\frac{1}{5}$

g. $\frac{2}{5}$ h. $\frac{3}{5}$

i. $\frac{4}{5}$ j. $\frac{5}{6}$

11. Design a procedure for adding two fractions with the same unit such as $\frac{2}{6}$ and $\frac{5}{6}$. Describe a rule. Check it out with at least two other examples.

add   numerators
denominators   the   same

12. Addition of fractional numbers was possible with white ones, but it is not so simple with many ones. (Give one counterexample to this conjecture.) Consider the arrangement in Figure 5. We may try to add two

fractional numbers by laying the left strips of each pair end to end and placing this chain to the left of the longest "one" strip. Is this the sum of the two fractional numbers?

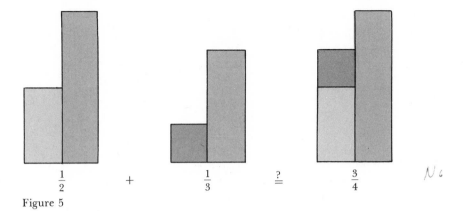

$\frac{1}{2}$ + $\frac{1}{3}$ $\overset{?}{=}$ $\frac{3}{4}$ *No*

Figure 5

13. Addition of white ones worked with the simple end-to-end operation. It doesn't work with many ones. Design a procedure for adding any two fractional numbers by manipulating their strip representations.

same   denominator

14. Use your procedure from problem 13 to find the sums of these pairs of fractions.

$\frac{3}{6} + \frac{4}{6} = \frac{7}{6}$

   a. $\frac{1}{2} + \frac{2}{3}$ b. $\frac{1}{3} + \frac{3}{4}$

   c. $\frac{2}{4} + \frac{2}{3}$ d. $\frac{1}{6} + \frac{1}{3}$

   e. $\frac{1}{5} + \frac{3}{10}$ f. $\frac{3}{5} + \frac{2}{3}$

15. Select a black strip and a light blue strip. Start making a train of black strips and a train of light blue strips side by side until they are the same length. This final common length is called the *least common multiple* (lcm) (page 159, Chapter 7). What is the least common multiple of these pairs of strips?

   a. White and rose   *12*

   b. Orange and light blue   *30*

   c. Purple and black   *14*

   d. Pink and white   *12*

16. One way to add fractions represented with strips is described in the following figures.

Represent the fractions (Figure 6).

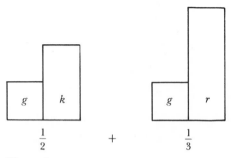

Figure 6

Find the lcm of the two denominators (Figure 7).

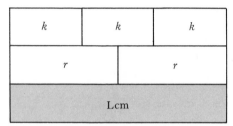

Figure 7

Represent each fraction as an equivalent fraction using the lcm as the new denominator (Figure 8).

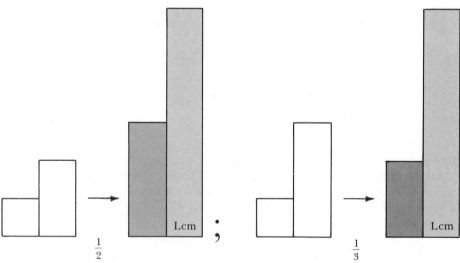

Figure 8

Now add these two fractions by chaining both numerators next to one of the denominators (Figure 9).

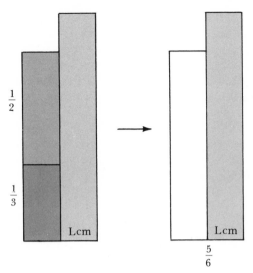

Figure 9

The result is $\frac{5}{6}$.

The steps for adding fractions this way are:

STEP 1 Represent the two fractions.

STEP 2 Find the lcm for the denominators.

STEP 3 Represent each fraction as an equivalent fraction using this common denominator.

STEP 4 Lay the two numerators end to end next to a common denominator.

STEP 5 Find the value of the new numerator.

STEP 6 Write the resulting fraction.

17. Use the procedure in problem 16 to find the sums for:

a. $\frac{1}{2} + \frac{1}{4} =$ $\frac{2}{4} + \frac{1}{4} = \frac{3}{4}$ b. $\frac{2}{3} + \frac{5}{6} =$ _____

c. $\frac{1}{5} + \frac{1}{2} =$ _____   d. $\frac{3}{4} + \frac{2}{3} =$ _____

e. $\frac{3}{4} + \frac{3}{8} =$ _____   f. $\frac{2}{6} + \frac{2}{3} =$ _____

g. $\frac{1}{3} + \frac{1}{2} + \frac{1}{4} =$ _____   h. $\frac{1}{5} + \frac{1}{4} =$ _____

18. As you have probably already conjectured, it would be impossible to make up a table for all pairs of fractions. There are just too many. But you can answer these questions with some well-founded conjectures.

a. Is addition closed on the set of fractions?

yes

b. Is addition associative on the set of fractions?

yes

c. Is there an additive identity element in the set of fractions?

0

d. Are there additive inverses in the set of fractions?

yes - if negatve

e. Is addition commutative on the set of fractions?

yes

19. Design a subtractive procedure for fractions represented by the strips.

common   Denominator

20. Find the difference between the two fractions of each pair:

a. $\frac{3}{4} - \frac{1}{4} =$ _____   b. $\frac{5}{6} - \frac{1}{3} =$ _____

c. $\frac{7}{8} - \frac{1}{2} =$ _____   d. $\frac{3}{5} - \frac{1}{4} =$ _____

e. $\frac{2}{3} - \frac{1}{2} =$ _____   f. $\frac{9}{10} - \frac{2}{5} =$ _____

g. $\frac{11}{12} - \frac{2}{3} =$ _____   h. $\frac{1}{2} - \frac{3}{4} =$ _____

21. Check these properties for subtraction of fractions.

a. Subtraction on the set of fractions is closed.

yes

b. Subtraction on the set of fractions is associative.

no

c. Subtraction on the set of fractions has an identity element.

no

d. There are subtractive inverses for every fraction in the set.

no

e. Subtraction on the set of fractions is commutative.

no

22. Represent the two fractions $\frac{3}{4}$ and $\frac{2}{3}$ with arrangements of Figure 10.

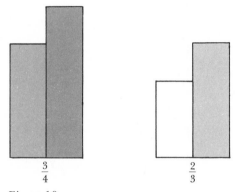

$\frac{3}{4}$ $\frac{2}{3}$

Figure 10

Take the two numerators and stack 'em up, then stretch 'em out as in Figure 11.

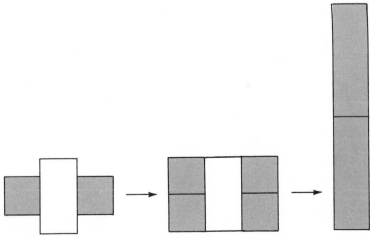

Figure 11

Do the same with the denominators (Figure 12).

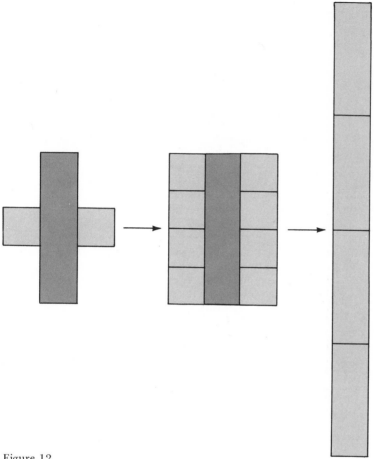

Figure 12

Now set them together to form a new fraction (Figure 13). Name this fraction and write its numeral.

$1/2$

23. Treat these pairs of fractions in the same way.

a. $\frac{1}{2}$; $\frac{2}{3}$: _____ 2/6

b. $\frac{2}{5}$; $\frac{1}{3}$: _____ 2/15

c. $\frac{1}{4}$; $\frac{1}{2}$: _____ 1/8

d. $\frac{2}{4}$; $\frac{1}{6}$: _____ 2/24

e. $\frac{5}{6}$; $\frac{1}{3}$: _____ 5/18

f. $\frac{1}{5}$; $\frac{3}{4}$: _____ 3/20

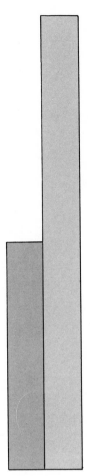

Figure 13

24. The process used in problem 22 is a model for the multiplication of fractions. Note that the usual sentences for multiplication can be written as follows:

a. $\frac{1}{2} \times \frac{2}{3} = \frac{2}{6}$ (or $\frac{1}{3}$)   b. $\frac{2}{5} \times \frac{1}{3} = \frac{2}{15}$

c. $\frac{1}{4} \times \frac{1}{2} = \frac{1}{8}$ d. $\frac{2}{4} \times \frac{1}{6} = \frac{2}{24}$ (or $\frac{1}{12}$)

e. $\frac{5}{6} \times \frac{1}{3} = \frac{5}{18}$ f. $\frac{1}{5} \times \frac{3}{4} = \frac{3}{20}$

Now complete this generalization: The product of the two fractions a/b and c/d is found by:

$$\frac{a}{b} \times \frac{c}{d} = \frac{ac}{bd}$$

25. Use this generalization to find the products of these pairs of fractions.

    a. $\frac{3}{8} \times \frac{1}{4} =$ $\frac{3}{32}$ b. $\frac{3}{4} \times \frac{1}{2} =$ _____

    c. $\frac{5}{8} \times \frac{1}{3} =$ _____   d. $\frac{2}{3} \times \frac{1}{6} =$ _____

    e. $\frac{1}{5} \times \frac{7}{8} =$ _____   f. $\frac{3}{5} \times \frac{5}{6} =$ _____

26. Check the following properties for multiplying fractions.

    a. Closure

    yes

    b. Associativity

    yes

    c. Identity

    yes $\frac{1}{1}$

    d. Inverses

    yes

    e. Commutativity

    yes

27. Represent the fraction $\frac{4}{5}$ with the white and light blue strips. Note that the white represents $\frac{4}{5}$ and the light blue denotes the unit. What is $\frac{1}{2}$ of $\frac{4}{5}$? Find $\frac{1}{2}$ of these fractions.

    $\frac{1}{2} \times \frac{4}{5} = \frac{4}{10} = \frac{2}{5}$

    a. $\frac{6}{10}$ $\frac{6}{10} \times \frac{1}{2} = \frac{6}{20} = \frac{3}{10}$

    b. $\frac{4}{7}$ $\frac{1}{2} \times \frac{4}{7} = \frac{4}{14} = \frac{2}{7}$

    c. $\frac{2}{3}$

    d. $\frac{5}{6}$

28. We usually represent one-half of a number by writing a division statement, such as

    $\frac{4}{5} \div 2 = \frac{2}{5}$

    Write such statements for each of the exercises in problem 27.

29. Complete these exercises.

    a. $\frac{6}{8} \div 3 = \frac{6 \times \frac{1}{3}}{8} = \frac{1}{4}$ (handwritten: $\frac{6 \times 1}{8 \quad 3} \div \frac{1}{4}$, $\frac{4}{10} \times \frac{1}{4} = \frac{1}{10}$)   b. $\frac{6}{8} \times \frac{1}{3} = $ _____

    c. $\frac{4}{10} \div 4 = \underline{\frac{4}{10} \times \frac{1}{4} = \frac{1}{10}}$ d. $\frac{4}{10} \times \frac{1}{4} = $ _____

    e. $\frac{8}{15} \div 2 = $ _____   f. $\frac{8}{15} \times \frac{1}{2} = $ _____

    g. $\frac{15}{100} \div 5 = $ _____   h. $\frac{15}{100} \times \frac{1}{5} = $ _____

30. Explore more examples like the ones in problem 29, and write a careful conjecture about a relationship between multiplication and division of fractions.

    $\frac{a}{b} \div \frac{c}{d} = \frac{a}{b} \times \frac{d}{c}$

31. Try the conjecture of problem 30 on these exercises.

    a. $\frac{2}{5} \div \frac{2}{3} = \underline{3/5}$

    b. $\frac{4}{3} \div \frac{5}{2} = $ _____

    c. $\frac{6}{8} \div \frac{6}{3} = $ _____

    d. $\frac{10}{15} \div \frac{2}{5} = $ _____

32. Represent the fraction $\frac{10}{15}$ with a two-strip stack numerator (2×5) and a two-strip stack denominator (3×5). Note the "common" factor in both the numerator and denominator. By removing these common factors, we can simplify our representation from $\frac{10}{15}$ to $\frac{2}{3}$. Are these two numbers equivalent? Remove common factors to simplify these quotients.   *(handwritten: yes)*

    a. $\frac{2}{6} \div \frac{4}{3} = \underline{\frac{2}{6} \times \frac{3}{4}}$ b. $\frac{7}{10} \div \frac{7}{5} = $ _____

    c. $\frac{6}{8} \div \frac{9}{12} = $ _____   d. $\frac{2}{3} \div \frac{6}{4} = $ _____

    e. $\frac{1}{6} \div \frac{5}{10} = $ _____   f. $\frac{4}{3} \div \frac{2}{6} = $ _____

    g. $\frac{5}{8} \div \frac{10}{4} = $ _____   h. $\frac{1}{8} \div \frac{5}{16} = $ _____

33. You have found that dividing a number by 2 is the same as multiplying by $\frac{1}{2}$. What happens when you multiply these two numbers together? ($2 \times \frac{1}{2} = 1$). Likewise, dividing by 3 gives the same result as multiplying by $\frac{1}{3}$. What is the product of 3 and $\frac{1}{3}$? Since $\frac{1}{3} \times 3 = 1$ and 1 is the identity for multiplication, we say that $\frac{1}{3}$ is the *inverse under multiplication* of 3, or 3 is the inverse under multiplication of $\frac{1}{3}$. Find the inverses under multiplication of these numbers:

    a. $\frac{2}{3}$ b. $\frac{5}{4}$ $\frac{3}{2}$ $\frac{4}{5}$

    c. $\frac{4}{7}$ d. $\frac{2}{5}$ $\frac{7}{4}$

    e. $\frac{6}{5}$ f. $\frac{10}{3}$

    g. $\frac{5}{7}$ h. $\frac{5}{6}$

34. Inverses under multiplication are sometimes called *reciprocals* of each other. Which number does not have a reciprocal? Which number is its own reciprocal?

   0   1

35. A famous (infamous?) rule of elementary mathematics is: To divide two fractions, "invert and multiply." Devise a manipulative procedure with the colored strips to make this procedure more understandable. Describe your invention with sketches and detailed directions.

36. Complete the following series of operations. Notice that the divisors are getting smaller and smaller. Make up a rule about dividing by zero.

$100 \div 1000 = 1/10$

$100 \div 100 = 1$

$100 \div 10 =$ _ *10* _

$100 \div 1 =$ _ *100* _

$100 \div 1/10 =$ _ *1000* _

$100 \div 1/100 =$ *10000*

$100 \div 1/1000 =$ _____

$100 \div 1/100,000 =$ _____

$100 \div 1/1,000,000 =$ _____

$100 \div 1/10,000,000 =$ _____

$100 \div 0 =$ _ *∞* _

3. Decimals

The relationship between the black strip and the light blue strip can be expressed as $\frac{2}{5}$. We can also represent this relationship by dividing the light blue strip into the black strip in the following manner.

STEP 1 Represent the division of the black by the light blue as in Figure 14.

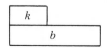

Figure 14

STEP 2 Since the black is shorter than the light blue, we will multiply the black by the dark blue and the light blue by the dark blue (Figure 15). Does this change the value of the fraction? **NO**   ⑩

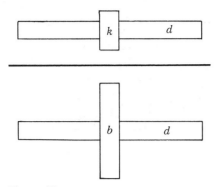

Figure 15

STEP 3 Find the chain for the black times the dark blue (Figure 16).

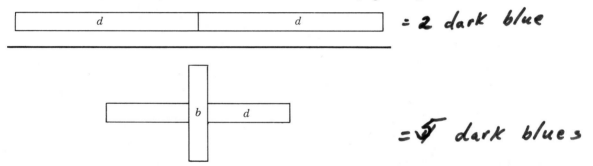

= **2 dark blue**

=**5 dark blues**

Figure 16

STEP 4 Now set the denominator dark blue strip aside and divide the numerator chain by the light blue (Figure 17).

d		d	
b	b	b	b

Figure 17

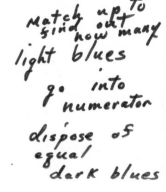

Match up to find out how many
4 light blues
go into numerator
dispose of equal dark blues

STEP 5 Record the number of light blues by writing the numeral 4. Move the decimal point one place to the left because the dark blue strip that is left in the denominator means we must divide by 10 (Figure 18).

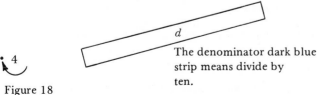

The denominator dark blue strip means divide by ten.

.4

Figure 18

Try this procedure with $\frac{3}{4}$.

STEP 1   Represent $\frac{3}{4}$ by a rose strip over a white strip (Figure 19).

Figure 19

STEP 2 Multiply the numerator and denominator by dark blue strips (Figure 20).

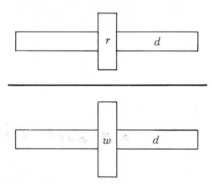

Figure 20

STEP 3 Divide the white into the numerator chain. It goes into the chain 7 times with a black strip remaining (Figure 21).

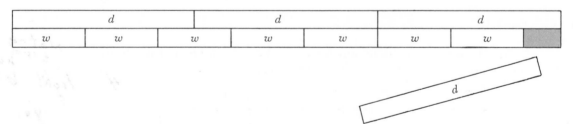

Figure 21

STEP 4 Record the 7 with a decimal point to the left since there is a ten strip in the denominator. This extra ten strip indicates that the 7 stands for seven-tenths.

STEP 5   Multiply the black strip by the dark blue strip and place another dark blue strip in the denominator with the white strip (Figure 22).

STEP 6 Now divide the white into this numerator chain (Figure 23).

STEP 7 Record the 5 to the right of the 7 (in the hundredths place) since there is a stack of ten strips in the denominator. Since there was no remainder in the division, we have an answer of 0.75 (Figure 24).

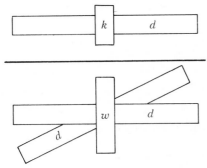

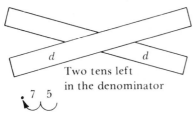

Figure 22

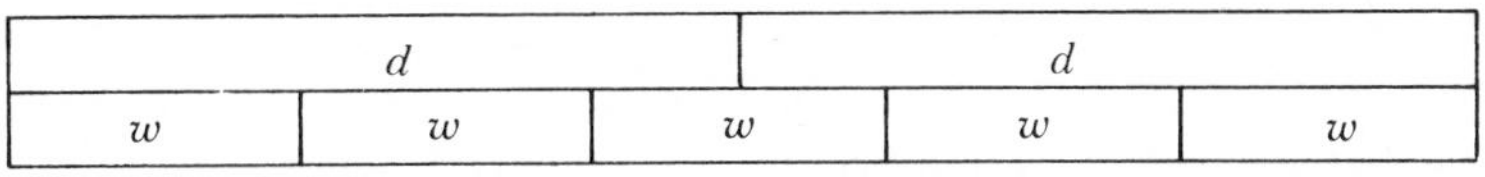

d			d	
w	w	w	w	w

Figure 23

d   d
Two tens left
in the denominator
.7 5

Figure 24

1. Using the procedure just outlined, convert the following fractions to decimals.

    a. $\frac{1}{4} =$ _.25_   b. $\frac{3}{5} =$ _.6_

    c. $\frac{3}{8} =$ _.375_   d. $\frac{2}{4} =$ _.5_

    e. $\frac{4}{5} =$ _.8_   f. $\frac{7}{8} =$ _.875_

    g. $\frac{3}{10} =$ _.5_   h. $\frac{3}{20} =$ _.15_

2. Reverse the process so that you can start with a decimal (say, 0.25) and get back to a fraction. Analyze the steps of the procedure described in this section and then apply them in reverse. Describe the steps with the strips.

STEP 1   Describe 0.25 with the strips.

STEP 2   (You continue.)

3. Convert these decimals to fractions.

   a. 0.5   b. 0.6

   c. 0.4   d. 0.3

   e. 0.125   f. 0.45

   g. 0.375   h. 0.85

4. Make a conjecture about the procedure for converting these decimals to their fractional equivalents.

*Place   The   given number   over,
The   respective place   value unit*

5. Some fractions are special in that they cannot be represented as a decimal in the usual way. Convert $\frac{1}{3}$ to a decimal. Make up a rule for handling this situation.

6. Convert these fractions to decimals.

   a. $\frac{2}{3}$

   b. $\frac{1}{6}$

   c. $\frac{5}{6}$

   d. $\frac{2}{7}$

7. The usual notation for repeating decimals is to place a bar over the digits that repeat. Here's how to write the decimals for problem 6:

   a. $\frac{2}{3} = 0.\overline{6}$

   b. $\frac{1}{6} = 0.1\overline{6}$

   c. $\frac{5}{6} = 0.8\overline{3}$

   d. $\frac{2}{7} = 0.\overline{285714}$

   To get these results, place zeros after the numerator and keep dividing by the denominator until the digits of the quotient repeat.

8. Convert these fractions to decimals.

   a. $\frac{1}{7} =$ _____   b. $\frac{4}{3} =$ _____

   c. $\frac{5}{9} =$ _____   d. $\frac{5}{7} =$ _____

   e. $\frac{8}{9} =$ _____   f. $\frac{7}{6} =$ _____

9. Another way to convert these fractions is to divide the denominator into the numerator with the usual division algorithm. Thus $\frac{3}{4}$ would be converted this way:

$$4\overline{)3}$$

$$4\overline{)3.0}$$

$$\begin{array}{r} .7 \\ 4\overline{)3.0} \end{array}$$

$$\begin{array}{r} .7 \\ 4\overline{)3.0} \\ \underline{2\ 8} \end{array}$$

$$\begin{array}{r} .7 \\ 4\overline{)3.0} \\ \underline{2\ 8} \\ 2 \end{array}$$

$$\begin{array}{r} .7 \\ 4\overline{)3.00} \\ \underline{2\ 8} \\ 20 \end{array}$$

$$\begin{array}{r} .75 \\ 4\overline{)3.00} \\ \underline{2\ 8} \\ 20 \end{array}$$

$$\begin{array}{r} .75 \\ 4\overline{)3.00} \\ \underline{2\ 8} \\ 20 \\ \underline{20} \\ 0 \end{array}$$

10. Beside each step of problem 9, describe the strip procedure we would use to get the same decimal.

11. Convert these fractions to decimals using the pencil-and-paper method:

a. $\frac{3}{5}$

b. $\frac{3}{4}$

c. $\frac{1}{4}$

d. $\frac{2}{5}$

e. $\frac{3}{8}$

f. $\frac{1}{6}$

g. $\frac{2}{3}$

h. $\frac{1}{7}$

12. Take all of the dark blue flats, dark blue strips, and green units and think of a way to represent these decimals. (*Hint:* If the flat is one-tenth, what is the strip? The unit?)

   a. 0.2   b. 0.31

   c. 0.024   d. 0.001

   e. 0.232   f. 0.601

13. Lay out the organizer sheet from the appendix. Represent the addition problem

     0.234
   + 0.310

as in Figure 25. Collect the strips, flats, and units, and record the sum.

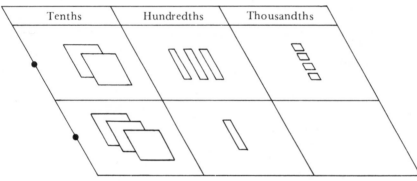

| Tenths | Hundredths | Thousandths |

Figure 25

Now lay the dark blue flats, dark blue strips, and unit strips in appropriate places on the organizer for these addition problems.

   a.   0.023   b.   0.302
       + 0.315   + 0.626

   c.   0.059   d.   0.090
       + 0.623   + 0.666

   e.   0.398   f.   0.2
       + 0.267   + 0.63

14. Note that you can also manipulate the flats, strips, and units in a way that is analogous to the standard algorithm for subtraction of decimals. Do the following subtraction problems using the dark blue flats, strips, and

units on the organizer sheet. Be sure to think of the second number as being "taken away" from the first number.

a.   0.636   b.   0.529
   − 0.327   − 0.016

c.   0.468   d.   0.999
   − 0.402   − 0.899

e.   0.600   f.   0.902
   − 0.260   − 0.666

15. The product 7×5 is 35. The product 7×0.5 is 3.5 (half of 7). The product 0.7×0.5 is 0.35 (half of $\frac{7}{10}$). To calculate such products involving decimals, we need to devise a system of dealing with the placement of the decimal point. The product 0.7×0.5 can be found in the following way:

$$0.7 \times 0.5 = \tfrac{7}{10} \times \tfrac{5}{10} = \frac{35}{10 \times 10} = \frac{3.5}{10} = 0.35$$

This procedure converts the two-decimal fractions to ratios with a unit (denominator) that is a multiple of 10. Note that the numerator is a whole number. Now we can simply multiply the numerators as whole numbers, then move the decimal point one place to the left for each factor of 10 that is in the denominator. Try this procedure on these products.

$$\textit{EXAMPLE} \quad 0.35 \times 0.3 = \frac{35}{10 \times 10} \times \frac{3}{10} = \frac{105}{10 \times 10 \times 10} = 0.105$$

 a. $0.2 \times 0.3 =$

 b. $0.9 \times 0.1 =$

 c. $0.21 \times 0.5 =$

 d. $0.45 \times 0.25 =$

 e. $0.02 \times 0.05 =$

 f. $0.005 \times 0.041 =$

16. Of course, the procedure described in Figure 26 is the one you have used before in finding products of decimal fractions. Calculate the following products and think about how your procedure is like the one described in Figure 26.

$$\begin{array}{r} .64 \\ \times .25 \\ \hline \end{array} \longrightarrow \begin{array}{r} 64 \div (10 \times 10) \\ 25 \div (10 \times 10) \\ \hline \end{array} \longrightarrow \begin{array}{r} 64 \div (10 \times 10) \\ 25 \div (10 \times 10) \\ \hline 320 \\ 128 \\ \hline 1600 \div (10 \times 10 \times 10 \times 10) \end{array} \longrightarrow \begin{array}{r} 64 \div (10 \times 10) \\ 25 \div (10 \times 10) \\ \hline 320 \\ 128 \\ \hline 1600 \end{array}$$

Figure 26

a.
$$\begin{array}{r} 0.23 \\ \times\ 0.02 \\ \hline \end{array}$$

b.
$$\begin{array}{r} 0.05 \\ \times\ 0.81 \\ \hline \end{array}$$

c.
$$\begin{array}{r} 0.31 \\ \times\ 0.68 \\ \hline \end{array}$$

d.
$$\begin{array}{r} 0.025 \\ \times\ 0.008 \\ \hline \end{array}$$

e.
$$\begin{array}{r} 0.301 \\ \times\ 0.662 \\ \hline \end{array}$$

f.
$$\begin{array}{r} 0.352 \\ \times\ 0.801 \\ \hline \end{array}$$

17. The division of 240 by 0.15 can be expressed as $\frac{240}{0.15}$, which is equivalent to

$$\frac{240}{0.15} \times 1$$

But we can use the name $\frac{100}{100}$ for 1 and write

$$\frac{240}{0.15} \times \frac{100}{100} = \frac{24000}{15}$$

which can be readily calculated as 1600. Use a similar procedure for these exercises.

a. $\dfrac{64}{0.2}$

b. $\dfrac{15}{0.03}$

c. $\dfrac{6}{0.0012}$

d. $\dfrac{51}{0.00017}$

18. The division of 0.036 by 12 can be expressed as:

$$\frac{0.036}{12} = \frac{0.036}{12} \times 1 = \frac{0.036}{12} \times \frac{1000}{1000}$$

$$= \frac{36}{12} \times \frac{1}{1000} = 3 \times \frac{1}{1000} = \frac{3}{1000} = 0.003$$

Try the procedure on these exercises.

a. $\dfrac{0.04}{8}$

b. $\dfrac{0.0025}{50}$

c. $\dfrac{0.0008}{20}$

d. $\dfrac{0.00017}{51}$

19. Use the ideas of problems 17 and 18 to explain this rule for dividing decimals using the usual algorithm:

$$0.23\overline{)342.608} = 23\overline{)34{,}260.8}$$

20. Divide 2,221.164 by 2.37 using the usual algorithm. Explain each step as you do the calculations.

$$2.37\overline{)2{,}221.164}$$

4. Percents

Percents are nothing but a standardized way of looking at relationships. They are much like "white ones" in that they use a standard base. But the base is much larger (100) than the white-ones system (where it is 4) and more useful in many ways. When we express the ratio 3/5 as 60/100, we use just the numeral 60 and call it 60%. Likewise, we can convert the ratio of 16 hits per 40 times at bat (16/40) to a fraction with a unit of 100 (40/100) and call this a 40% batting average. *Percent* means units per hundred.

1. Convert these fractions to percents.

a. 3/4   b. 7/20

c. 21/300   d. 33/75

e. 39/150   f. 27/90

g. 1/3   h. 2/7

2. Convert these percents to simple fractions.

a. 60%   b. 75%

c. 32%   d. 48%

e. 92%   f. 35%

g. 80%   h. 59%

3. Percentages are also useful in comparing ratios from very unlike situations. A 10% cut in a famous musician's income is considerably more than a 10% cut of your income. Yet both percents represent the same ratio of 10 to 100. Find the selling prices of these items with the indicated changes:

Original Price	Percent change	New Price
a. $3,000.00 small car	plus 10%	
b. $29,000.00 home	minus 12%	
c. $219.00 tent	plus 5%	
d. $89.00 suit	minus 15%	

4. Select a rubber band 10–20 cm long and about 0.5 cm (or 5 mm) wide. Cut and lay it flat next to a ruler. Mark off 11 equally spaced intervals at least 1 cm apart. Label them 0, 10, 20, 30, 40, 50, 60, 70, 80, 90, and 100. Note that the spaces are still equally spaced when you stretch the rubber band.

Now make a chain of 2 dark blue strips and another of 1 pink strip (Figure 27). Now stretch the rubber band along the longer chain (20 cm) with the zero mark at one end and the 100 mark at the other end. Now read the mark on the rubber band at the end of the pink strip. It should be close to 60. This means that the pink strip is 60% as long as the 2 dark blue strips. Use this same procedure to compare the following pairs of strips.

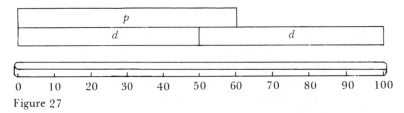

Figure 27

   a. purple with turquoise

   b. brown with pink

   c. orange with dark blue

   d. light blue with pink

5. Make chains for these pairs of numbers, and use the percent calculator your rubber band has become to determine the relationship of their lengths.

   a. 12 and 20

   b. 18 and 30

   c. 15 and 25

   d. 19 and 15

6. Make chains for these pairs of numbers, and use the rubber band percent calculator to determine the relationship of their lengths.

a. 14 and 27

b. 12 and 25

c. 17 and 21

d. 6 and 35

5. *Scientific notation*

Sometime you may have used scientific notation for dealing with very large or very small base-ten numbers. This notation is a simple procedure for expressing numbers in a standard form.

EXAMPLE $247 = 2.47 \times 10 \times 10$

EXAMPLE $0.0247 = 2.47 \times \frac{1}{10} \times \frac{1}{10}$

EXAMPLE $0.0000247 = 2.47 \times \frac{1}{10} \times \frac{1}{10} \times \frac{1}{10} \times \frac{1}{10} \times \frac{1}{10}$

1. Convert these decimals to a product of a number between 1 and 10, and ten to some power.

   a. 32   b. 4287

   c. 6,042,000   d. 0.421

   e. 0.0276   f. 0.000000027

   g. 247.621   h. 3,000,241.00002

2. As you found in problem 1, it is cumbersome to record the tens or one-tenths. A better procedure is to use an exponent system (as described in Chapter 5) to denote the number of factors. Complete the following sequence to determine a good rule for using exponents for this notation system.

EXAMPLE $100,000 = 10 \times 10 \times 10 \times 10 \times 10 = 10^5$

EXAMPLE $10,000 = 10 \times 10 \times 10 \times 10 \quad\quad = 10^4$

   a. 1,000 $= 10 \times 10 \times 10$ $= 10$

   b. 100 $= 10 \times 10$ $= 10$

   c. 10 $= 10$ $= 10$

   d. 1 $=$ $= 10$

   e. 0.1 $= \frac{1}{10}$ $= 10$

   f. 0.01 $= \frac{1}{10} \times \frac{1}{10}$ $= 10$

   g. 0.001 $= \frac{1}{10} \times \frac{1}{10} \times \frac{1}{10}$ $= 10$

   h. 0.0001 $= \frac{1}{10} \times \frac{1}{10} \times \frac{1}{10} \times \frac{1}{10}$ $= 10$

   i. 0.00001 $= \frac{1}{10} \times \frac{1}{10} \times \frac{1}{10} \times \frac{1}{10} \times \frac{1}{10}$ $= 10$

3. Using negative numbers as exponents for scientific notation of numbers less than 1 is an arbitrary choice that may or may not be meaningful to you. It turns out that it is a very useful choice when more complex manipulations of fractions are required.

EXAMPLE $\dfrac{100,000}{100} = \dfrac{10^5}{10^2} = 10^3$

EXAMPLE $\dfrac{1,000}{1,000,000} = \dfrac{10^3}{10^7} = 10^{-4}$

a. $\dfrac{100}{10,000}$

b. $\dfrac{10}{100,000}$

c. $\dfrac{100,000}{100}$

d. $\dfrac{1,000,000}{10,000}$

4. Now convert the following numbers to standard scientific notation. You should have a number that is equal to or greater than 1, but less than 10, multiplied by 10 to some power.

EXAMPLE $3,789 = 3.789 \times 10^3$

EXAMPLE $0.0215 = 2.15 \times 10^{-2}$

a. 243   b. 3,428

c. 9,098,000   d. 0.024

e. 0.00438   f. 0.0000009

g. 243.09   h. 0.00000000006

6. *Names for fractions in base five*

The cleverness and simplicity of decimal fractions is difficult to appreciate because they are so familiar. To develop some appreciation and insight into naming fractions in a base and place-value numeration system, try the following exercises.

1. If the number 1 is represented by a collection of five light blue flats, then one light blue flat would be one-fifth. Under these same conditions, what is the value of the light blue strip? The green unit?

2. The collection of light blues and greens in Figure 28 would be named 0.243*b* according to this system. Name each of the collections of light blues and greens with numerals.

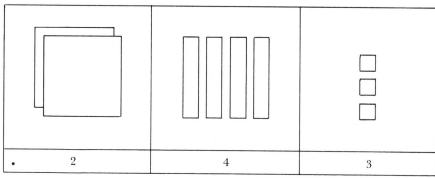

Figure 28

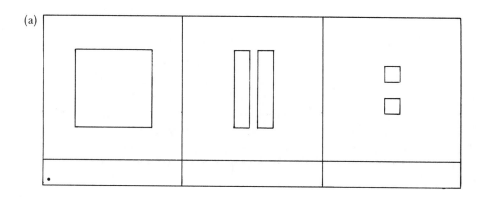

(a)

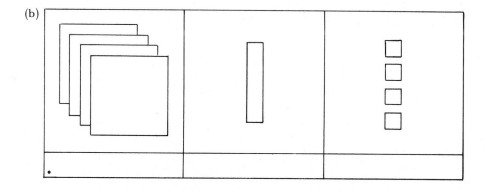

(b)

(c)

(d)

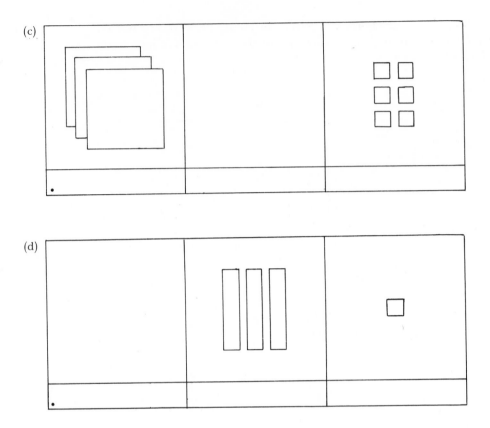

3. To talk about these base-five names for fractions, we shall introduce words to use for the numbers. The light blue flats will be called "quinths," the light blue strips will be called "quarths," and the green squares will be called "qubths." Then the names for the fractions would use the system set forth in Figure 29. This particular fraction's name is "2 quinths, 3 quarths, one qubth."

Quinths	Quarths	Qubths

Figure 29

Name these base-five fractions after representing them with standard light blue.

a. 0.412*b*

b. 0.024*b*

c. 0.304*b*

d. 0.230*b*

4. Use the organizer sheet in the appendix for organizing the light blue flats, light blue strips, and green squares for the following exercises.

a.	0.231*b* + 0.113*b*	b.	0.041*b* + 0.302*b*
c.	0.024*b* + 0.241*b*	d.	0.001*b* + 0.004*b*
e.	0.432*b* − 0.021*b*	f.	0.342*b* − 0.121*b*
g.	0.340*b* − 0.122*b*	h.	0.140*b* − 0.044*b*

7. *Fractional numbers on a geoboard*

A fractional number can be associated with the ordered pair of whole numbers that gives the location of each peg on the geoboard relative to the peg in the lower left corner. This lower left corner peg is called the *origin*.

Fractional numbers are represented on the geoboard in the following manner: The denominator is the number of units to move *right* from the origin, and the numerator is the number of units to move *up* from the origin (or, in practice, from the end of the denominator).

EXAMPLE   The fractional number $\frac{1}{2}$ is at X in Figure 30. X is 2 units right (the denominator) and 1 unit up (the numerator) from the origin.

1. What fractional numbers are represented at the following points in Figure 30?

A _____   B _____   C _____

2. Represent the following fractional numbers on your geoboard and record them in Figure 30.

a. $\frac{3}{5}$

b. $\frac{5}{6}$

c. $\frac{4}{7}$

d. $\frac{4}{3}$

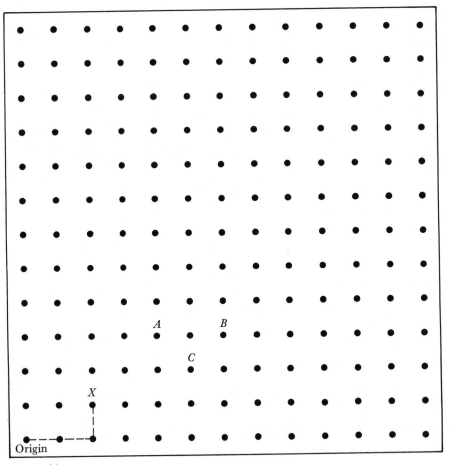

Figure 30

3. Graph $\frac{1}{2}, \frac{4}{8}, \frac{3}{6}, \frac{2}{4}$ on your geoboard.

   a. What can you say about their graphic representations?

   all on sTraight Line

   b. How are these numbers related?

   Same ratio

   c. Place a rubber band around the origin and the representative of $\frac{4}{8}$.

d. How is the band related to the other points?

passes through all points

e. How would the representations of $\frac{5}{10}$ and $\frac{6}{12}$ relate to all these points?

Same line

4. Graph $\frac{2}{3}$, $\frac{4}{6}$, $\frac{6}{9}$ on your geoboard and record them in Figure 31.

a. How are these fractions related?

equal

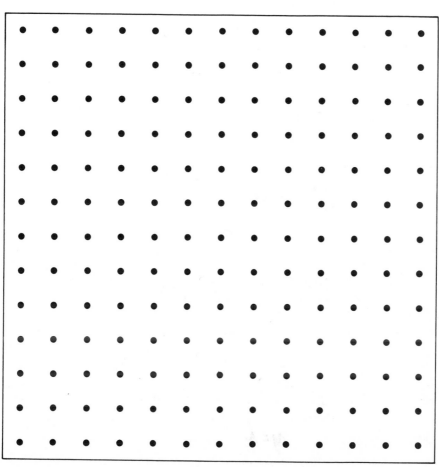

Figure 31

b. Would a line drawn through these three points pass through the origin?

yes

5. Make a conjecture about the graphic representations for this type of fractional number.

st. Line.

6. Graph $\frac{1}{2}$ on your geoboard. Beginning at the representation of $\frac{1}{2}$, graph $\frac{1}{2}$ again. Continue the pattern in Figure 32 on your geoboard. Make a conjecture about the fractional numbers in this pattern.

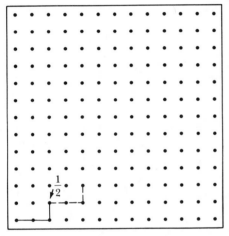

=

Figure 32

7. Graph on your geoboard the following fractional numbers and two numbers equivalent to each fraction, and record them in Figure 32.

a. $\frac{1}{4}$

b. $\frac{3}{4}$

c. $\frac{1}{3}$

8. The number 1 has many names. Place a rubber band around the representations of 1 on your geoboard and record them in Figure 33. This is called the *1-line*.

y=1

9. In the same manner, locate the "0-line" on your geoboard and record several names for zero.

x axis

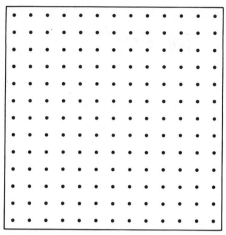

Figure 33

10. Place a rubber band around the representation for $\frac{3}{4}$ and the origin and another band around the $\frac{2}{3}$ and the origin as in Figure 34.

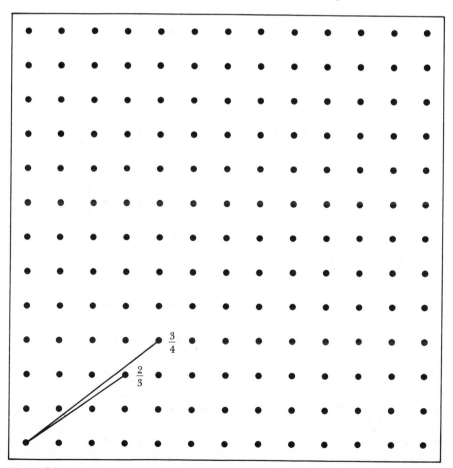

Figure 34

a. Make a conjecture about the size of these fractional numbers and the slope of the bands.

$$\frac{3}{4} > \frac{2}{3} \qquad 3/4 \ \text{steeper}$$

b. Based on your conjecture, determine on your geoboard which member of the following pairs of fractional numbers is greater. Circle the larger of the two numbers.

    i. $\frac{1}{2}$ and $\frac{2}{3}$ $\frac{2}{3}$

    ii. $\frac{4}{5}$ and $\frac{3}{4}$ $4/5$

    iii. $\frac{3}{5}$ and $\frac{4}{7}$ $3/5$

11. The fractional numbers $\frac{1}{4}$ and $\frac{2}{4}$ have like denominators. To add $\frac{1}{4}$ and $\frac{2}{4}$, add their distances above the 0-line. One way is to begin at $\frac{1}{4}$ and move up 2 units; another is to begin at $\frac{2}{4}$ and move up 1 unit (Figure 35).

Add the following fractional numbers on your geoboard and record the sum in Figure 35.

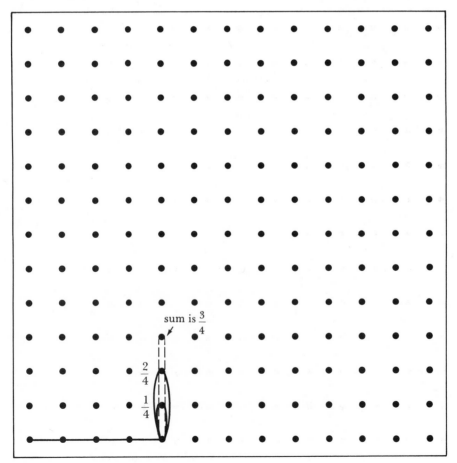

Figure 35

a. $\frac{2}{5} + \frac{4}{5}$

b. $\frac{1}{3} + \frac{4}{3}$

c. $\frac{2}{6} + \frac{3}{6}$

12. Does the order of addition make a difference in the sum?

13. Graph $\frac{1}{2}$ and $\frac{1}{3}$ on your geoboard. Since these graphs are not in the same column, they cannot be added. However, if equivalent fractions with like denominators can be found, *they* can be added instead.

Find the column that contains both a representation of a fraction equivalent to $\frac{1}{2}$ and a graph of a fraction equivalent to $\frac{1}{3}$. What is $\frac{1}{2} + \frac{1}{3}$?

$$\frac{1}{6} \; , \; \frac{1}{2} + \frac{1}{3} = \frac{5}{6}$$

14. Solve these problems on your geoboard and record your results in Figure 36.

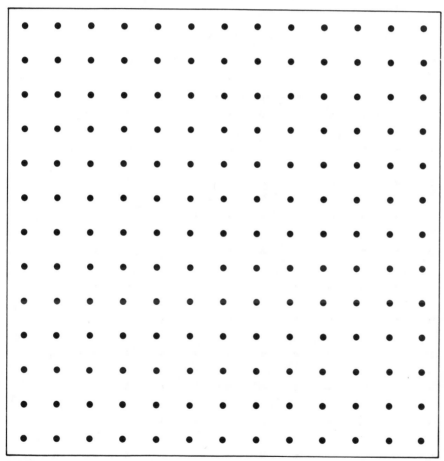

Figure 36

a. $\frac{2}{3} + \frac{1}{6}$

b. $\frac{1}{3} + \frac{2}{2}$

c. $\frac{3}{4} + \frac{1}{2}$

Only fractions with like denominators can be subtracted. Consequently, equivalent fractions may have to be used, just as we used them to add fractional numbers. The solution of $\frac{1}{2} - \frac{1}{3}$ is the difference between the numerators or heights of their graphs, once they are written as equivalent fractions with like denominators.

15. Graph $\frac{1}{2}$ and $\frac{1}{3}$ again on your geoboard and find their difference.

16. To find $\frac{2}{3} \times \frac{3}{4}$, graph $\frac{3}{4}$ on your geoboard. Make the 4-denominator three times as long and the 3-numerator two times as long to locate the peg of the product of $\frac{2}{3}$ and $\frac{3}{4}$.

   a. Is there a simpler name for the product?

   b. Explore whether the order of multiplication makes a difference.

17. Find the following products on your geoboard. Record the answers in Figure 37.

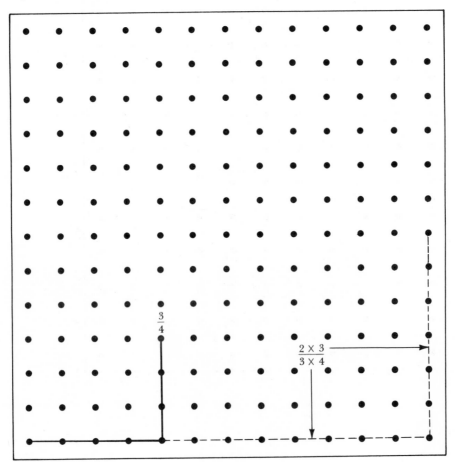

Figure 37

a. $\frac{1}{3} \times \frac{3}{4}$

b. $\frac{1}{2} \times \frac{3}{5}$

c. $\frac{1}{2} \times \frac{3}{4}$

Remember, two fractions whose product is 1 are called *multiplicative inverses*.

EXAMPLE $\frac{2}{3} \times \frac{3}{2} = 1$, so $\frac{3}{2}$ is the multiplicative inverse of $\frac{2}{3}$.

18. Place a rubber band around the graph of $\frac{2}{5}$ and the graph of its multiplicative inverse, and compare the band with the graph of the 1-line.

   a. Try this with $\frac{1}{4}$, $\frac{3}{7}$, and $\frac{2}{6}$.

   b. Make a conjecture about fractional numbers and their multiplicative inverses in relation to the 1-line.

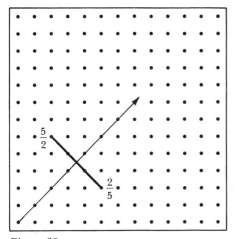

Perpendicular To 1 line.

Figure 38

Division is the inverse operation of multiplication, so $\frac{2}{3} \div \frac{3}{4}$ is the same as $\frac{2}{3}$ times the multiplicative inverse of $\frac{3}{4}$. So instead of finding the quotient $\frac{2}{3} \div \frac{3}{4}$, we find the product $\frac{2}{3} \times \frac{4}{3}$, which is $\frac{8}{9}$.

19. Solve the following on your geoboard and record the solution in Figure 39.

   a. $\frac{1}{2} \div \frac{2}{3}$

   b. $\frac{3}{4} \div \frac{1}{2}$

   c. $\frac{2}{4} \div \frac{3}{2}$

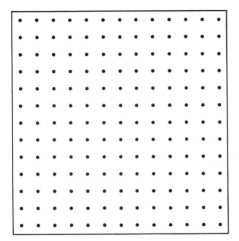

Figure 39

Summary

Rational numbers are more difficult to understand and use than whole numbers partly because of the variety of ways in which each number can be represented. Furthermore, each number can represent several different concepts. In the time you have had for this chapter, it was impossible to fully develop the concept of rational numbers. No doubt you realize that you use decimal representations of rationals most often and find them more familiar.

With metric measurement becoming the new measurement standard in the United States, even more decimal notation will be used. The use of traditional fractions will slowly diminish to occasional application of basic fractions such as $\frac{1}{2}$ and $\frac{2}{3}$. In the meantime you should be able to deal with fractions and the operations of addition, subtraction, multiplication, and division in both systems. Although this may seem demanding to us, we really have no choice. We happen to live at that time in history when the conversion to metric measurement is taking place. The best we can do is try to thoroughly understand the basis for fractions and the two ways of expressing them.

References

Brumfiel, Charles. "Mathematical Systems and Their Relationships to the Real World." *The Arithmetic Teacher,* 17 (November 1970), 563–573.

O'Brien, Thomas. "Two Approaches to the Algorithm for Multiplication of Fractional Numbers." *The Arithmetic Teacher,* 12 (November 1965), 552–555.

Smith, Frank. "Divisibility Rules for the First Fifteen Primes." *The Arithmetic Teacher,* 18 (February 1971), 85–87.

Winzenread, Marvin. "Repeating Decimals." *The Arithmetic Teacher,* 20 (December 1973), 679–682.

10

Number sentences

Algebra is a versatile tool, and the skills required to apply it can be useful in many practical situations. Unfortunately, most people do not feel comfortable enough with algebra to use it with accuracy and confidence. The concepts and procedures of algebra have been developed by mathematicians in a precise, deductive way. For our purposes it is more appropriate to use an informal, inductive approach.

We chose the number balance as one useful means of efficiently producing physical situations that can be easily converted into algebraic language. The arithmetic in this chapter is intentionally not difficult. We want you to be able to spend your time carefully observing the development of the concepts and skills. It is especially important that you not just get answers, but look instead for the relationships between what is happening with the number balance and the recording you do in symbolic and graphic form.

1. The balance

Remove the balance stand and arm from the insert at the back of the book and construct it according to the directions in the appendix. Select about 20 paper clips that are very close to the same weight. Design a procedure that could be used to select paper clips of almost identical weights.

1. Place a paper clip on the number 5 on the left side, a paper clip on number 3 on the left side, and a paper clip on number 8 on the right side as in Figure 1. What happens?

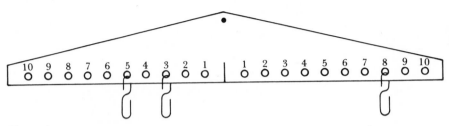

Figure 1

2. Place the following arrangements of paper clips on the number balance and record whether they balance, they do not balance, or you are not able to tell.

a. A paper clip on 6 left, 4 left, and 10 right.   ~~BALANCED~~ NOT BALANCED   UNKNOWN

b. A paper clip on 5 left, 3 right, and 4 right.   BALANCED (NOT BALANCED)   UNKNOWN

c. A paper clip on the right side, 6 left, and 3 right.   BALANCED NOT BALANCED (UNKNOWN)

d. A paper clip on 8 left, 4 left, 10 right, and 2 right.   (BALANCED) NOT BALANCED   UNKNOWN

e. A paper clip on 7 left, 2 left, and a paper clip on the right side.   BALANCED NOT BALANCED (UNKNOWN)

f. A paper clip on 10 left, 2 right, and 5 right.   BALANCED (NOT BALANCED) UNKNOWN

3. a. Make a conjecture on the arrangements that balanced in problem 2.

  Sum of left = Sum of Right

b. Make up some arrangements of clips on the balance to check your conjecture.

4. Place 3 paper clips on 3 on the left side and a single paper clip on 9 on the right side, as in Figure 2. What happens?

  Balances

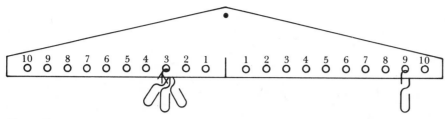

Figure 2

5. Place the following arrangements on paper clips on the number balance and record whether they balance, they do not balance, or you are not able to tell.

a. 2 paper clips on 4 left and 1 paper clip on 8 right.   (BALANCED)   NOT BALANCED   UNKNOWN

b. 1 paper clip on 10 left and 5 paper clips on the right side.   BALANCED   NOT BALANCED   (UNKNOWN)

c. 3 paper clips on 5 left and 1 paper clip on 8 right.   BALANCED   (NOT BALANCED)   UNKNOWN

d. 4 paper clips on the left side and 2 paper clips on 10 right.   BALANCED   NOT BALANCED   (UNKNOWN)

e. 3 paper clips on 5 left, 1 paper clip on 10 right, and 1 paper clip on 5 right.   (BALANCED)   NOT BALANCED   UNKNOWN

f. A paper clip on 7 left and 2 paper clips on 5 right.   BALANCED   (NOT BALANCED)   UNKNOWN

6. a. Make a conjecture about the arrangements that balanced.

sum of the products = sum of products

b. Make up some arrangements to check your conjecture.

7. Based on your conjectures in problems 3 and 6, write a number sentence that describes each of the following balance pictures. Note that for a sentence to be an equation, the clips must balance. Consequently, the = sign is used to indicate that the balance arm is level.

EXAMPLE

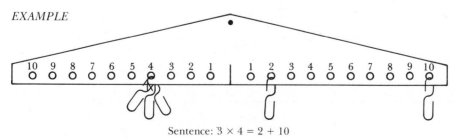

Sentence: $3 \times 4 = 2 + 10$

Figure 3

a. Sentence:

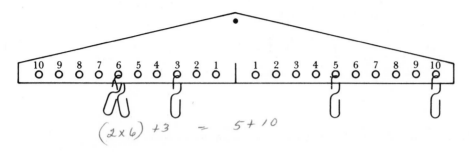

$(2 \times 6) + 3 \quad = \quad 5 + 10$

b. Sentence:

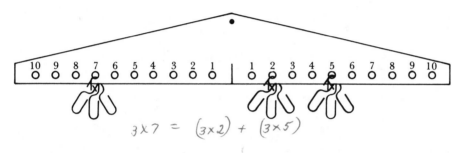

$3 \times 7 \quad = \quad (3 \times 2) + (3 \times 5)$

c. Sentence:

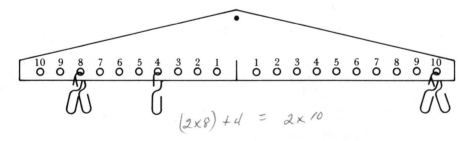

$(2 \times 8) + 4 \quad = \quad 2 \times 10$

d. Sentence:

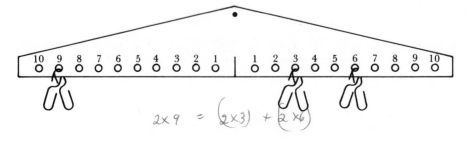

$2 \times 9 \quad = \quad (2 \times 3) + (2 \times 6)$

8. Place the following arrangements on the number balance and write a sentence that describes the relationship between the numbers of paper clips and their position on the balance.

EXAMPLE   A paper clip on each of 4 left, 6 left, and 10 right.

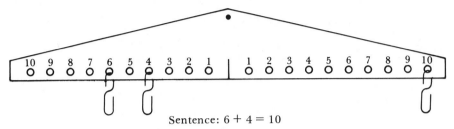

Sentence: $6 + 4 = 10$

Figure 4

a. Three paper clips on 6 left, 1 paper clip on 10 right, and 1 paper clip on 8 right.

Sentence: $3 \times 6 = 1 \times 10 + 1 \times 8$ or $3 \times 6 = 10 + 8$

b. Two paper clips on 3 left, 2 paper clips on 5 left, and 2 paper clips on 8 right.

Sentence: $(2 \times 3) + (2 \times 5) = 2 \times 8$

c. Two paper clips on 10 left, 1 paper clip on 2 left, 4 paper clips on 5 right, and 1 paper clip on 2 right.

Sentence: $(2 \times 10) + 2 = (4 \times 5) + 2$

d. One paper clip on 9 left, 1 paper clip on 6 left, 1 paper clip on 5 left and 2 paper clips on 10 right.

Sentence: $9 + 6 + 5 = 2 \times 10$

9. Represent the following number sentences with paper clips on the balance, and record the arrangement by sketching the paper clips.

a. $3 \times 5 = 15$

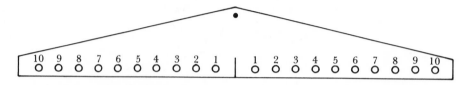

b. $13 = 4 + 9$

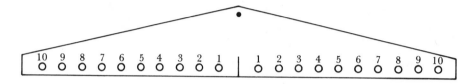

c. $(3 \times 6) + 4 = 22$

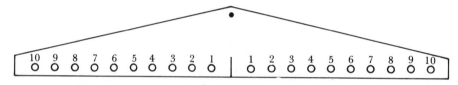

d. $2 \times 8 = 9 + 7$

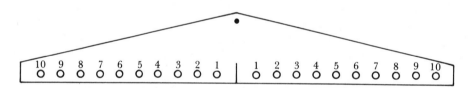

Sentences involving the equals (=) relation that balance are called *true*, and those that do not balance are called *false*.

10. Represent the following number statements on the balance and determine whether they are true or false.

a. $3 \times 4 = 6 + 6$ (TRUE)   FALSE

b. $9 + 3 = 16$ TRUE   (FALSE)

c. $5 + (2 \times 3) = 9$ TRUE   (FALSE)

d. $(3 \times 4) + (3 \times 6) = 3 \times 10$ (TRUE)   FALSE

e. $9 + 7 = 10 + 8$ TRUE   (FALSE)

Those sentences for which you cannot tell what the relationship is are called *open sentences*. An open box or some other shape is often used to designate the information missing in an open sentence. This shape is called a *placeholder*.

11. Place the following arrangements on the number balance, and write an open sentence for each of the arrangements.

EXAMPLE Three paper clips on the left side, 1 paper clip on 10 right, and 1 paper clip on 8 right.

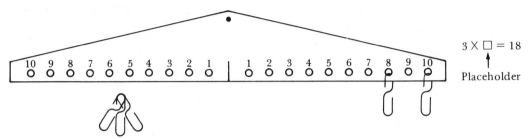

$3 \times \square = 18$
↑
Placeholder

Figure 5

a. One paper clip on 10 left, 1 paper clip on 6 left, and 2 paper clips on the right side.

$$10 + 6 = 2 \times \square$$

b. Four paper clips on the left side, 1 paper clip on 2 left, and 3 paper clips on 10 right.

$$(4 \times \square) + 2 = 3 \times 10$$

c. Two paper clips on 5 left, 1 paper clip on the left, 1 paper clip on 10 right, and 1 paper clip on 4 right.

$$(2 \times 5) + \square = 10 + 4$$

d. One paper clip on 8 left, 1 paper clip on 6 left, and 2 paper clips on the right side.

$$8 + 6 = 2 \times \square$$

12. Represent the following open sentences on the balance. In the placeholder, record the head of an arrow pointing toward the side that tips up on the balance.

EXAMPLE $5 + 6 \boxed{>} 9$

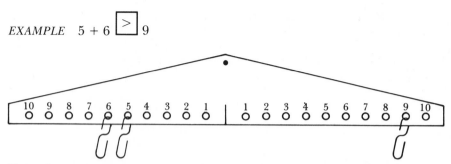

Figure 6

a. $(2 \times 8) + 3 \boxed{<} 21$ b. $17 \boxed{>} 9 + 5$

c. $3 \times 5 \boxed{<} 9 + 8$ d. $26 \boxed{>} 4 \times 6$

e. $(2 \times 4) + 6 \boxed{<} 10 + 7$

13. If the left side tips up as in part (c) of problem 12, the balance indicates that the left side is *less than* the right side. Consequently, we shall use the head of an arrow pointing to the left as the symbol for *less than* relation. The sentence $3 \times 5 < 9 + 8$ then means that "three times five is less than nine plus eight."

If the right side tips up as in part (b) of problem 12, we can see on the balance that the left side is *greater than* the right side. The head of an arrow pointing to the right then shall be used as the symbol for the *greater than* relation. Consequently, $17 > 9 + 5$ means that "seventeen is greater than nine plus five."

Represent the following number sentences on the balance, and indicate whether they are true, false, or open by circling the appropriate word.

EXAMPLE $2 + \boxed{} = 5$ TRUE   FALSE   (OPEN)

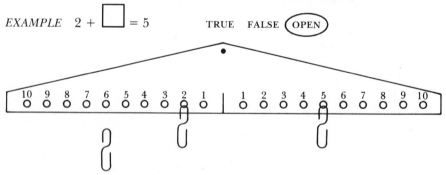

Figure 7

a. $16 + 6 = 10$ (TRUE) (FALSE) OPEN

b. $13 + 6 < 18$ TRUE   (FALSE) OPEN

c. $9 \bigcirc 7 > 20$ TRUE   FALSE   (OPEN)

d. $16 + 5 = 21$ (TRUE) FALSE   OPEN

e. $\left(3 \times \boxed{}\right) \times 4 = 13$ TRUE   FALSE   (OPEN)

f. $5 \times 9 = (5 \times 2) + (5 \times 7)$ TRUE   (FALSE)   OPEN

g. $12 > 7 + 9$ TRUE   (FALSE)   OPEN

h. $3 \times 7 \,\triangle\, 10$ TRUE   FALSE   (OPEN)

2. Appropriate replacements

If one open sentence contains several placeholders of the same shape (a box, for example), then whatever replacement you use in the first of these placeholders, you must also put in all the others of the same shape.

EXAMPLE $\square \times \square = 16$

If 2 is placed in the $\square$, then the resulting statement is $2 \times 2 = 16$, which is false. Furthermore,

$$\boxed{2} \times \boxed{8} = 16$$

is *incorrect* according to the rule of substituting, even though $2 \times 8 = 16$ is a true sentence.

EXAMPLE $\boxed{7} + \boxed{3}\!\triangle = \boxed{3}\!\triangle + \boxed{7}$

This statement is correct according to the rule for substituting because the same number is in both boxes and the same number is in both triangles.

EXAMPLE $\left(2 \times \boxed{2}\right) + \boxed{2}\!\triangle = 6$

These are appropriate replacements that make the resulting sentence true. Note that the replacement may be the same for different placeholders within a sentence.

1. Identify those sentences that have appropriate replacements in the placeholders.

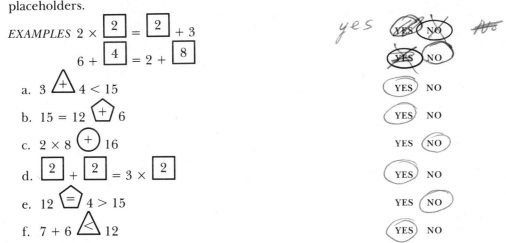

EXAMPLES $2 \times \boxed{2} = \boxed{2} + 3$ *yes*   (YES) (NO)   *no*

         $6 + \boxed{4} = 2 + \boxed{8}$ (YES) NO

a. $3 \,\triangle\!\!+\,\, 4 < 15$ (YES)   NO

b. $15 = 12 \,\pentagon\!\!+\,\, 6$ (YES)   NO

c. $2 \times 8 \,\bigcirc\!\!+\,\, 16$ YES   (NO)

d. $\boxed{2} + \boxed{2} = 3 \times \boxed{2}$ (YES)   NO

e. $12 \,\pentagon\!\!=\,\, 4 > 15$ YES   (NO)

f. $7 + 6 \,\triangle\!\!<\,\, 12$ (YES)   NO

2. Replace the placeholders with appropriate numerals, operational signs, or relation symbols; then determine if the resulting statements are true or false.

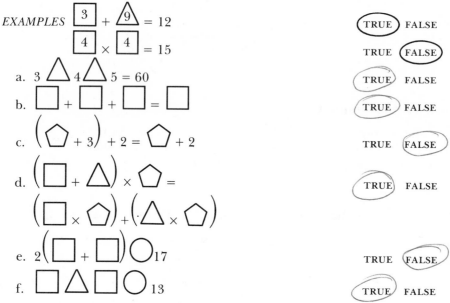

EXAMPLES $\boxed{3} + \triangle{9} = 12$ (TRUE) FALSE

$\boxed{4} \times \boxed{4} = 15$ TRUE (FALSE)

a. $3 \triangle 4 \triangle 5 = 60$ (TRUE) FALSE

b. $\square + \square + \square = \square$ (TRUE) FALSE

c. $\left(\pentagon + 3 \right) + 2 = \pentagon + 2$ TRUE (FALSE)

d. $\left(\square + \triangle \right) \times \pentagon =$ (TRUE) FALSE

$\left(\square \times \pentagon \right) + \left(\triangle \times \pentagon \right)$

e. $2\left(\square + \square \right) \bigcirc 17$ TRUE (FALSE)

f. $\square \triangle \square \bigcirc 13$ (TRUE) FALSE

3. Make replacements in the following open sentences so that the statements are false.

a. $8 + 7 \bigcirc{>} 15$

b. $4 \times \boxed{2} = 4 \times \left(3 + 4 \right)$

c. $4 + 3 \pentagon{<} 7$

d. $9 \times 2 \boxed{\neq} 18$

e. $12 + \triangle{3} > 16$

4. The set of those replacements that make an open sentence true is called the *truth set*. The balance may be used to find the truth set for open sentences. Find the truth set for the following open sentences on the balance and shade it in on the balance pictures.

EXAMPLE $3 + \square = 5$

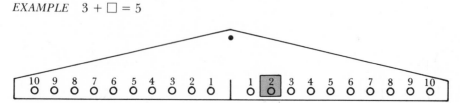

Figure 8

EXAMPLE $5 < \square + 3$

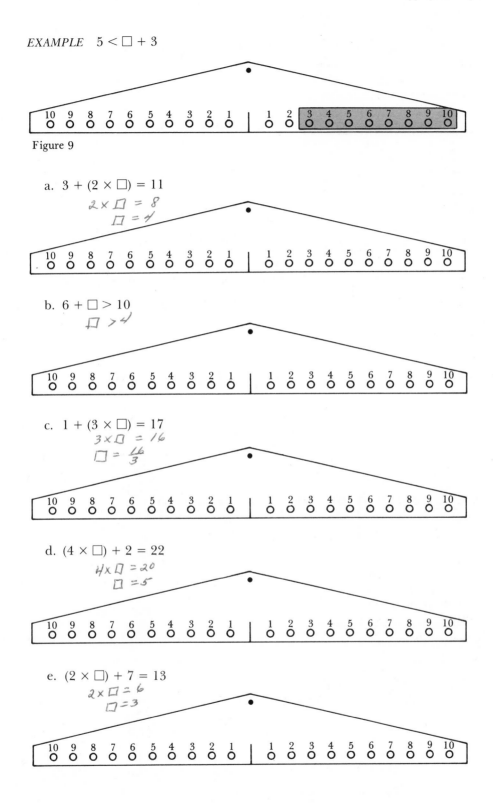

Figure 9

a. $3 + (2 \times \square) = 11$

$2 \times \square = 8$
$\square = 4$

b. $6 + \square > 10$

$\square > 4$

c. $1 + (3 \times \square) = 17$

$3 \times \square = 16$
$\square = \frac{16}{3}$

d. $(4 \times \square) + 2 = 22$

$4 \times \square = 20$
$\square = 5$

e. $(2 \times \square) + 7 = 13$

$2 \times \square = 6$
$\square = 3$

f. $12 < 3 \times \square$

$4 < \square$

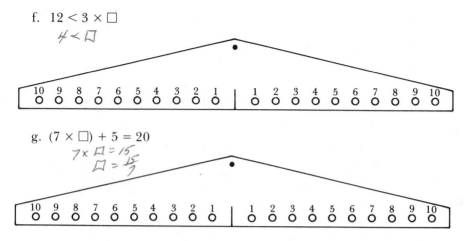

g. $(7 \times \square) + 5 = 20$

$7 \times \square = 15$

$\square = \dfrac{15}{7}$

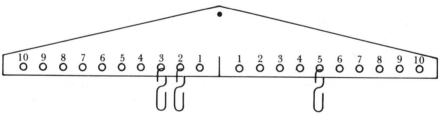

5. Make conjectures about the nature of the graphs of the truth set for an open sentence involving the following relations: $=, <, >$.

$=$ one   solution

$<$ and $>$ many   solutions

6. Find the truth set for the following sentences by finding the appropriate operation or relation signs.

EXAMPLE $3 \boxed{+} 2 = 5$

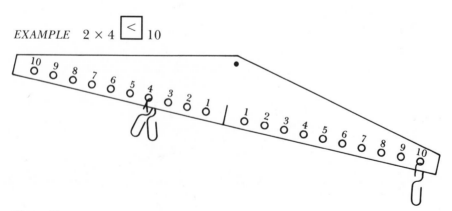

Figure 10

EXAMPLE $2 \times 4 \boxed{<} 10$

Figure 11

a. 6 ⊠ 3 = 9 △+ 9

b. 8 ⬭> (2 + 3) ☐− 1

c. 9 ⬠⊠ 8 > 20

d. 12 ⊠ 7 > 22

7. Determine whether the following open sentences are true for all replacements, some replacements, or no replacements.

a. $4 + \square < \square + 4$ ALL   SOME   (NONE)

b. $8 \times \square = \square$ ALL   (SOME)   NONE

c. $\square + \square = 2 \times \square$ (ALL)   SOME   NONE

d. $\square \times \square < 1$ ALL   (SOME)   NONE

e. $4 + (3 \times \square) = (3 \times \square) + 4$ (ALL)   SOME   NONE

f. $\square \times \square = \square + \square$ ALL   (SOME)   NONE

g. $(12 + \square) \times 1 = 12 \times \square$ ALL   SOME   (NONE)

h. $8 + \square = 8 \times \square$ ALL   SOME   (NONE)

i. $(4 \times \square) + (5 \times \square) = 9 \times \square$ (ALL)   SOME   NONE

j. $(6 \times \square) + 2 = 8$ ALL   (SOME)   NONE

8. Write four open sentences that are true for *all* replacements. Check them on your balance.

a. $\square = \square$

b.

c.

d.

9. Write four open sentences that are true for *some* replacements.

a. $\square + 3 = 5$

b.

c.

d.

10. Write four open sentences for which there are *no* replacements that will make them true.

a. $6 + \square > \square + 6$

b.

c.

d.

3. Simplifying clip arrangements

The sentence $(4 \times 4) + 2 = (2 \times 4) + 10$ is represented in (a) balance of Figure 12. If you remove 2 paper clips from 4 left, you may remove 2 paper clips from 4 right to make it balance. The resulting sentence, $(2 \times 4) + 2 = 10$, is represented by balance (b) of Figure 12.

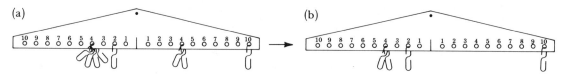

Figure 12

1. Represent $(2 \times 3) + 4 = 6 + 4$ on the (a) balance. Remove 1 clip from 4 right. Remove 1 clip from 4 left. Sketch the result on the (b) balance. Write the resulting sentence.

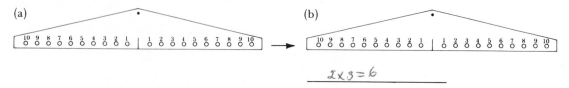

$2 \times 3 = 6$

2. Write a sentence for the arrangement on the (a) balance. Now remove 1 clip from 6 left and move the clip from 10 right to 4 right. Draw the resulting sketch on the (b) balance and write a corresponding sentence.

(a)   (b)

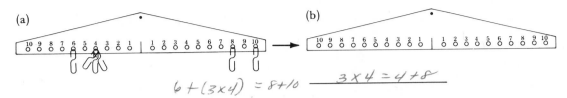

$$6 + (3 \times 4) = 8 + 10 \qquad \underline{3 \times 4 = 4 + 8}$$

3. Sketch the arrangement for $(5 \times 6) + 8 = 38$ on the (a) balance. Remove the paper clip from 8 left. What do you have to do to the right side to make it balance? Sketch the result on the (b) balance and write a corresponding sentence.

(a)   (b)

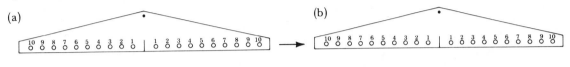

_____

4. Make a conjecture based on the previous three problems. Use sketches and corresponding sentences on the balance to check your conjecture.

(a)   (b)

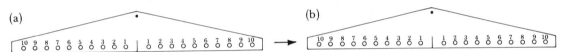

5. Place the following arrangements on the balance and write the appropriate number sentence for each arrangement.

EXAMPLE Three paper clips on 9 left, 3 paper clips on 6 right, and 3 paper clips on 3 right.

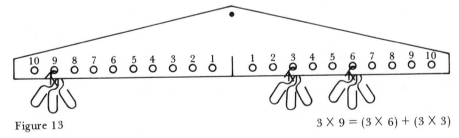

Figure 13 $3 \times 9 = (3 \times 6) + (3 \times 3)$

a. Four paper clips on 7 left, 4 paper clips on 5 right, and 4 paper clips on 2 right.

$$4 \times 7 = (4 \times 2) + 4 \times 5$$

b. Two paper clips on 8 left, 2 paper clips on 2 left, and 2 paper clips on 10 right.

$$(2 \times 8) + (2 \times 2) = 2 \times 10$$

c. Three paper clips on 6 left, 3 paper clips on 2 right, and 3 paper clips on 4 right.

$$3 \times 6 = (3 \times 2) + (3 \times 4)$$

d. Five paper clips on 6 left, 5 paper clips on 2 left, and 5 paper clips on 8 right.

$$(5 \times 6) + (5 \times 2) + 5 \times 8$$

e. Make a conjecture about the number sentences represented in the above exercises.

$$ab + ac = a(b + c)$$

6. Based on your conjecture, find the truth set for these open sentences.

a. $(3 \times 39) + (3 \times 11) = 3 \times \square$ 50

b. $6 \times 120 = (6 \times \triangle) + (6 \times 20)$ 100

c. $(5 \times \square) + (5 \times \triangle) = 5 \times 62$ 60 2

d. $\diamond \times 20 = (\diamond \times 13) + (\diamond \times 7)$ 8 *any number*

7. The sentence $3 \times 8 = (3 \times 6) + (3 \times 2)$ is represented on the (a) balance. Remove 1 paper clip from each position. Does it still balance? Sketch the resulting arrangement and write the corresponding sentence.

yes

(a)   (b)

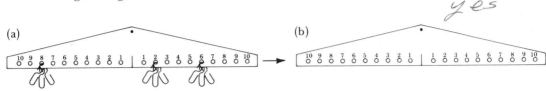

_____

8. Represent the sentence $4 \times 7 = (4 \times 2) + (4 \times 5)$ on the (a) balance. Remove half of the paper clips from each position. Does it still balance? Record the results on the (b) balance and write the corresponding sentence.

(a)   *yes*   (b)

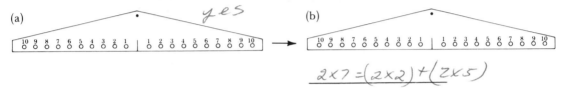

$2 \times 7 = (2 \times 2) + (2 \times 5)$

9. Write a sentence for the arrangement on the (a) balance. Remove 1 paper clip from each position. Record the results on the (b) balance and write the corresponding sentence.

(a)   (b)

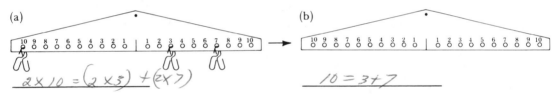

$2 \times 10 = (2 \times 3) + (2 \times 7)$ $10 = 3 + 7$

10. The sentence $(5 \times 6) + (5 \times 3) = 5 \times 9$ is represented on the (a) balance. Remove $\frac{4}{5}$ of the clips from each position. Sketch the result on the (b) balance. Write the resulting sentence.

(a)   (b)

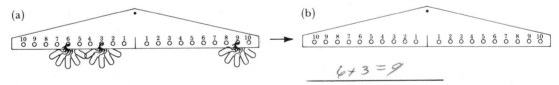

$6 + 3 = 9$

11. Represent the sentence $(6 \times 7) + (6 \times 2) = 6 \times 9$ on the (a) balance. Remove $\frac{2}{3}$ of the clips from each position. Sketch the result on the (b) balance. Write the corresponding sentence.

(a)   (b)

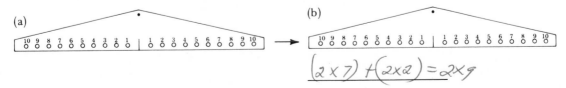

$(2 \times 7) + (2 \times 2) = 2 \times 9$

12. Make a conjecture about the last five exercises. Use the balance to demonstrate your conjecture.

(a)

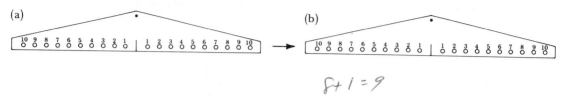

(b)

$$8 + 1 = 9$$

13. Transform these sentences into simpler ones.

EXAMPLE $(3 \times \square) + 6 = 30 \rightarrow 3 \times \square = 24$

EXAMPLE $(3 \times \square) + 6 = 30 \rightarrow \square + 2 = 10$

   a. $(8 \times \square) + 10 = 50$

   b. $(4 \times \square) + 8 = 32$

   c. $41 = (4 \times \square) + 9$

   d. $5 \times \square = (2 \times \square) + 15$

   e. $(3 \times \square) + 7 = \square + 23$

14. The main objective of transforming equations is to find an equivalent equation that is easier to solve. For instance, the equation $3 \times \square + 7 = 22$ could be transformed as follows: Subtract 7 from each side; divide both sides by 3. The resulting sentence is $\square = 5$. Check this out on the balance as in Figure 14.

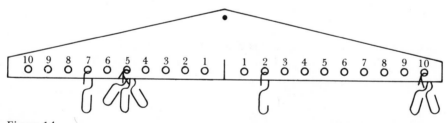

Figure 14

Replacing $\square$ with 5 balances so that 5 is a solution to the open sentence. The sequence of steps can be represented as $3 \times \square + 7 = 22$, then $3 \times \square = 15$, and finally $\square = 5$. How is each of these steps demonstrated on the balance?

15. Write out a sequence of steps that leads to a solution for each of the following open sentences. Check your answers on your balance.

   a. $(4 \times \square) + 5 = 33$ *7*

   b. $(\square \times 6) + (\square \times 4) + 3 = 33$ *3*

   c. $41 = (7 \times \square) + 6$ *5*

   d. $(7 \times \square) + (7 \times \triangle) + 5 = 40$ *3 5*

16. In the sentence $(3 \times \square) + 7 = 19$, subtracting 7 from both sides can be represented as $(3 \times \square) + 7 - 7 = 19 - 7$. An equivalent sentence is $3 \times \square = 19 - 7$. Consequently, $(3 \times \square) + 7 = 19$ may be transformed to $3 \times \square = 19 - 7$ by subtracting 7 from both sides. Likewise, $(7 \times \square) + 5 = 26$ may be transformed to $7 \times \square = 26 - 5$. Make similar transformations for these equations.

   a. $(5 \times \triangle) + 9 = 34$ $5 \times \triangle = 34 - 9$

   b. $(6 \times \square) + 12 = 24$ $6 \times \square = 24 - 12$

   c. $(12 \times \hexagon) + 4 = 40$ $12 \times \triangle = 40 - 4$

   d. $(9 \times \square) - 2 = 25$ $9 \times \square = 25 + 2$

17. We can do similar transformations with equations involving products. In the sentence $3 \times \square = 12$, we can divide each side by 3 to get $(3 \times \square) \div 3 = 12 \div 3$ which simplifies to $\square = 4$. Check this transformation: $6 \times \square = 18$; $(6 \times \square) \div 6 = 18 \div 6$; $\square = 3$. Write out transformations for these equations.

   a. $7 \times \hexagon = 21$ 3

   b. $4 \times \square = 28$ 7

   c. $(5 \times \triangle) + 7 = 22$ 3

   d. $(6 \times \square) - 5 = 13$ 3

   e. $3 \times (\square + 2) = 18$ 4

   f. $(\square \times 7) + (\square \times 3) = 110$ 11

18. Solve the following equations using the general rule of doing the same thing to both sides.

EXAMPLE $\left(2 \times \boxed{5}\right) + 7 = 17$

a. $\left(3 \times \boxed{}\right) = 6$ *2*

b. $\left(5 \times \boxed{}\right) + 15 = 40$ *5*

c. $\left(9 \times \boxed{}\right) + 5 = 59$ *6*

d. $\left(7 \times \boxed{}\right) + 3 = \left(5 \times \boxed{}\right) + 43$ *20*

e. $\left(6 \times \boxed{}\right) + 18 = 48$ *5*

f. $\left(100 \times \boxed{}\right) + 291 = 791$ *5*

g. $\left(8 \times \boxed{}\right) + 115 = \left(5 \times \boxed{}\right) + 127$ *4*

h. $\left(100 \times \boxed{}\right) + 21 = \left(99 \times \boxed{}\right) + 23$ *2*

19. Determine whether the rules for simplifying equations also hold for inequalities. That is, are $(3 \times \square) + 3 < 15$ and $3 \times \square < 12$ equivalent? And are $3 \times \square < 12$ and $\square < 4$ equivalent? Write your conclusions with supporting evidence. Give examples using the balance.

20. The symbols $\neq$, $\nless$, and $\ngtr$ represent the relations *not equal to, not less than*, and *not greater than*, respectively. How do your rules work with these relations? Give examples to justify your claims. Give examples using the balance.

4. Conventions for order of operations

Grouping symbols are frequently used in simplifying or clarifying number sentences containing three or more numerals. In general, we perform operations within the grouping symbols first. If symbols are *nested*, we perform operations *from inside out.*

1. Evaluate each of these number sentences.

EXAMPLE $\left[63 \div (7 + 2)\right] \times 5 = \left[63 \div 9\right] \times 5 = \left[7\right] \times 5 = \boxed{35}$

   a. $\left[(25 - 7) \div 6\right] \times 9 = \square$ *27*

   b. $43 - \left[(5 + 9) \times 3\right] = \square$ *1*

   c. $\left[8 \times (9 - 3)\right] \div 3 = \square$ *16*

   d. $\left[35 - (5 \times 6)\right] - 2 = \square$ *3*

   e. $\left[56 \div (5 + 3)\right] - 7 = \square$ *0*

   f. $\left[(7 + 9) \times 2\right] \div 4 = \square$ *8*

2. Insert parentheses ( ) and brackets [ ] in the following number sentences to make each a true sentence.

EXAMPLE $\left[24 \div (3 + 1)\right] - 5 = 1$

   a. $(24 \div 3) - (1 + 5) = 2$

   b. $(7 - 3) \times 4 - 5 = 11$

   c. $15 - [3 \times (3 + 2)] = 0$

   d. $27 - [(19 + 7) \div 13] = 25$

   e. $10 - [(7 + 3) \div 2] = 5$

   f. $[63 \div (7 + 2)] \times 4 = 28$

   g. $[18 \div (7 + 2)] \times 3 = 6$

   h. $(11 - 7) \times (36 \div 12) = 12$

   i. $[3 + (6 \times 8)] - (4 \div 4) = 50$

   j. $3 + 6 \times (8 - 4) \div 4 = 9$

3. To prevent confusion in evaluating an expression such as $6 + 4 \times 3$, mathematicians have agreed on certain steps to follow in simplifying expressions that do not contain grouping symbols. If we multiply first, the expression becomes $6 + 12$, or 18. If we add first, we arrive at 10×3, or 30. To avoid such ambiguity, we agree on these rules for order of operations when there are no parentheses.

First perform all multiplications and divisions in left-to-right order.

Second perform all additions and subtractions in left-to-right order.

EXAMPLE $17 - 9 \div 3 \times 4 = \boxed{5}$

Evaluate each of these number sentences.

a. $18 - \left[9 \div 9 \times 4\right] = \boxed{}$

b. $13 + 9 - \left(24 \div 3\right) = \boxed{}$

c. $\left(4 \times 6\right) - \left(21 \div 3\right) = \boxed{}$

d. $18 + \left(14 \div 2\right) - \left(3 \times 4\right) = \boxed{}$

e. $\left(6 \times 9\right) - \left(63 \div 9\right) + 5 = \boxed{}$

4. Use as few grouping symbols as possible to make the following sentences true.

EXAMPLE $6 \times 2 + 4 \times (3 + 2) = 32$

a. $\left(9 + 5\right) \div 7 \times 6 = 12$

b. $\left(7 \times 4 + 4\right) \div 4 = 8$

c. $\left(16 - 8\right) \times 4 \div 16 = 2$

d. $8 + 7 \times \left(27 \div 3 - 4\right) = 43$

e. $60 \div \left[\left(9 - 6\right) \times 5\right] + 7 = 11$

Summary

The rules for simplifying number sentences or solving open sentences are easily modeled on the number balance. The model provides a direct link between the abstract world of algebra and the real world of physics. The basic idea of distance from the fulcrum times units of weight is complex, yet it is easily demonstrable on the balance.

By now you are probably thinking that all elementary mathematics can be modeled with a structural device. It turns out that this is true, and it is an important idea for you to appreciate. Perhaps mathematics can be a reasonable and friendly approach to dealing with quantitative situations! The rules are not arbitrary but rather come from real situations. Once you know the rules, you can apply them to new situations.

Using the rules with confidence still doesn't make the relationships between variables understandable. Consequently, we need to develop some graphing skills to add a visual dimension to the meaning of a formula. That will be the objective of the next chapter.

References

Bidwell, James. "Number Sentence Trios." *The Arithmetic Teacher,* 21 (February 1974), 150–152.

Hajek, Roy D. "A Rationale in the Use of Variables." *The Arithmetic Teacher,* 13 (November 1966), 546–548.

May, Frank B. "Three Problems of Using Equations in Elementary Arithmetic Programs." *The Arithmetic Teacher,* 11 (March 1964), 166–168.

Van Engen, Henry. "Why Use Frames in Arithmetic?" *The Arithmetic Teacher,* 13 (April 1966), 315–316.

11

Equations
and graphs

The graph as a visual representation of abstract number sentences is a natural extension of working with a number balance and it fits our philosophy on learning mathematics. Visual images help us understand the abstract nature of the concepts and skills we are emphasizing. Graphs give us a specific idea of how the variables of a sentence are related. Even seemingly complex data can become meaningful when viewed as a graph.

Certainly graphs have always been essential to the scientific community, but they are becoming increasingly important to everyone. We are bombarded by statistical information in the form of complex graphs in newspapers, investment prospectuses, insurance advertisements, and someone's latest solution to the economic woes of the world. A *graph* should be a simplified picture of relationships among data. From it, you should be able to draw conclusions. This chapter explains how graphs are evolved from one particular set of circumstances: paper clips on a number balance.

1. Truth sets

Hang a paper clip on 5 right. Now hang another paper clip anywhere on the right side and a paper clip on the left side so that they balance, as in Figure 1.

The open sentence $\Box + 5 = \triangle$ describes the arrangement of Figure 1. Find all possible combinations on the number balance that make this open

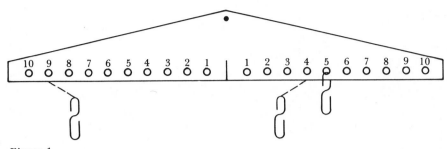

Figure 1

sentence true (i.e., balance). Record them in Figure 2. Each replacement must consist of a pair of numbers: one number to go into the box and one number to go into the triangle.

EXAMPLE $\boxed{4} + 5 = \triangle\!\!\!9$

$\square$	$\triangle$
4	9

Figure 2

When recording the pairs, we traditionally write the number that goes into the box first and the number that goes into the triangle second. Writing the pair as (4, 9) means that 4 goes in $\square$ and 9 goes in $\triangle$.

Replacing (4, 9) gives the true sentence

$\boxed{4} + 5 = \triangle\!\!\!9$

so the pair (4, 9) is a member of the truth set.

The order is important. The pair (9, 4) would mean that 9 goes in $\square$ and 4 goes in $\triangle$; this would produce the false sentence

$\boxed{9} + 5 = \triangle\!\!\!4$

1. Find on the balance four ordered pairs of the truth set for each of the following open sentences, and record them in the tables.

EXAMPLE $(2 \times \square) + 1 = \triangle$

$\square$	$\triangle$
1	3
2	5
3	7
4	9

a. $\square + 6 = \triangle$

$\square$	$\triangle$

b. $(3 \times \square) + 2 = \triangle$

$\square$	$\triangle$

c. $4 \times \square = \triangle$

$\square$	$\triangle$

d. $\triangle + \square = 10$

$\square$	$\triangle$

2. Members of the truth set for the open sentence in this chapter can be recorded on an array such as the geoboard in Figure 3. Call the dot in the lower left corner the *origin*. The first number of a pair (replacement for $\square$) is the number of units right from the origin, and the second number of the pair (replacement for $\triangle$) is the number of units up from the origin.

EXAMPLES   (3, 2) is recorded at A
        (2, 4) is recorded at B

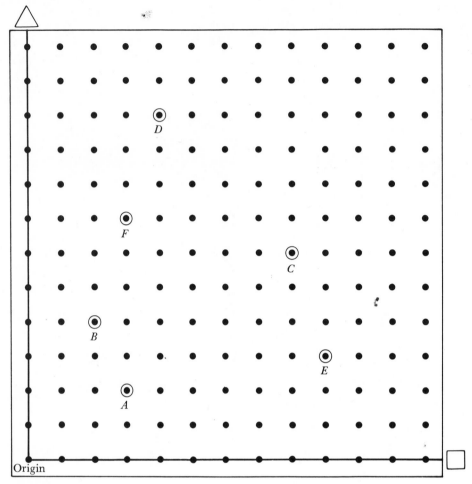

Figure 3

Record the following pairs in Figure 3 by circling the dots.

    a. (8, 2)

    b. (5, 7)

    c. (6, 6)

    d. (4, 1)

3. The following dots in Figure 3 are recordings of what pairs of numbers?

    C _____   D _____   E _____   F _____

4. Find on the balance several members of the truth set for the following open sentences, and record them on the arrays.

EXAMPLE   □ + 2 = △

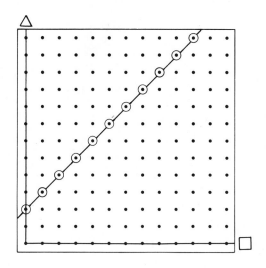

a. △ + □ = 8

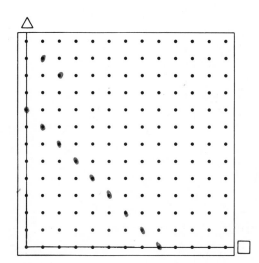

b. $(3 \times \square) + 5 = \triangle$

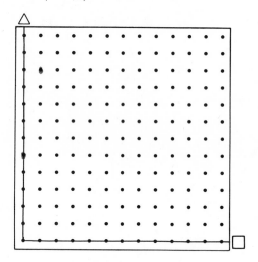

c. $4 \times \square = \triangle$

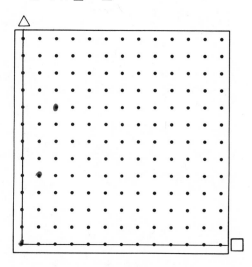

d. Make a conjecture about the graph of the truth set of open sentences like those in this problem.

straight Line

5. Find several members of the truth set of the following open sentences and record them in the arrays.

EXAMPLE $(2 \times \square) + 1 < \triangle$

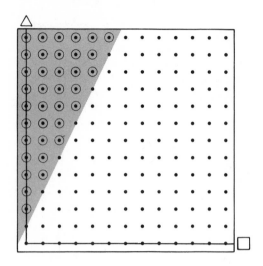

a. $(1/3 \times \square) + 2 > \triangle$

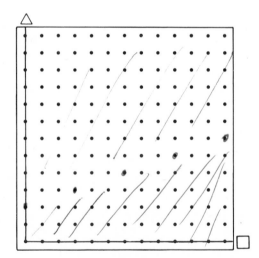

b. □ + 6 < △

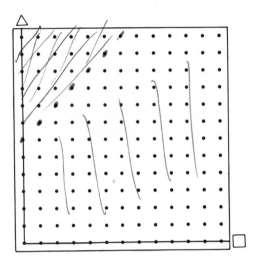

c. △ + (2 × □) < 9

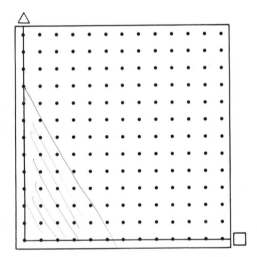

d. Make a conjecture about the graph of the truth set of open sentences like those of this problem.

plane

6. The graphs of the truth sets for three open sentences are given in (a), (b), and (c). Determine the members of the truth sets and record them in the appropriate tables.

a.

□	△
0	0
1	3
2	6
3	9
4	12

b.

□	△
0	5
3	6
6	7
9	8
12	9

c.

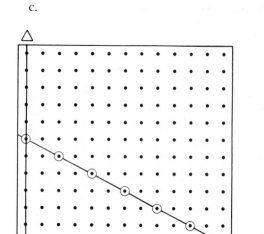

□	△
0	7
2	6
4	5
6	4

2. Graphs of open sentences

1. Continue the graphs for each of the open sentences without doing any arithmetic. Place the pairs you graphed in the open sentences and check them on the balance.

a. □ + 4 = △

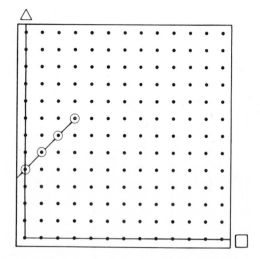

b. $(2/3 \times \square) + 3 = \triangle$

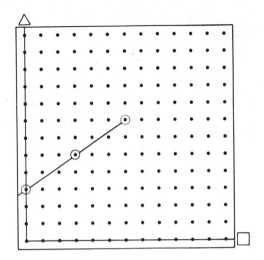

c. $(^-2 \times \square) + 9 = \triangle$

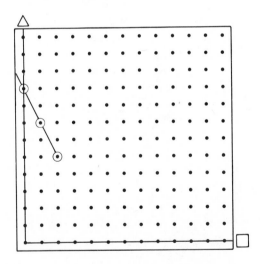

d. $(4 \times \square) + 1 = \triangle$

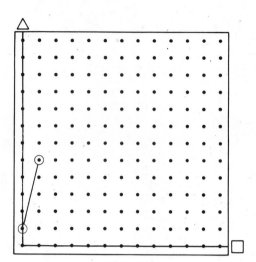

2. Describe the pattern you found in each of the graphs of the open
sentences in problem 1.

a. $\dfrac{1}{1}$

b. $\dfrac{2}{3}$

c. $-\dfrac{2}{1}$

d. $\dfrac{4}{1}$

3. The pattern in the graph of a truth set may be described as "1 unit to the
right, 4 units up," "3 units to the right, 2 units up," "1 unit to the right,
down 2 units," etc. This is called the *slope* pattern. The slope of a line graph
is the number of vertical units divided by the number of units to the right
from one peg to another. If the vertical move is up, the slope is positive (+);
if the vertical move is down, the slope is negative (−).

EXAMPLE $(3 \times \square) + 2 = \triangle$
            Pattern: right 1, up 3
            Slope $= \dfrac{3}{1}$

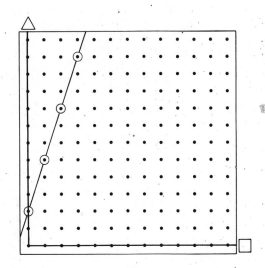

EXAMPLE $(^-2 \times \square) + 10 = \triangle$
            Pattern: right 1, down 2
            Slope $= \dfrac{^-2}{1}$

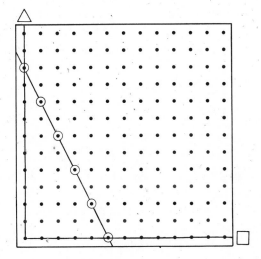

The following are open sentences and graphs of their truth sets. Determine the pattern in the graphs and compute the slope of each of the graphs.

a. $(4 \times \square) + 0 = \triangle$
   Pattern: $\frac{4}{1}$
   Slope = _____

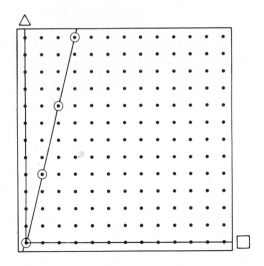

b. $(1/2 \times \square) + 5 = \triangle$
   Pattern: $\frac{1}{2}$
   Slope = _____

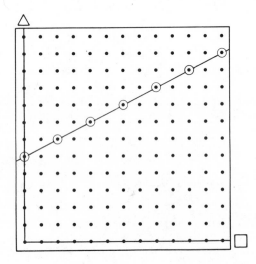

c. $(^-3 \times \square) + 9 = \triangle$
   Pattern:
   Slope = ___ $-\frac{3}{1}$ ___

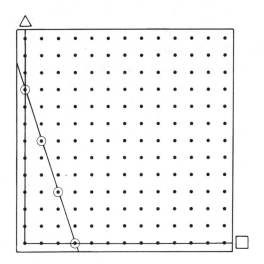

d. $(2/3 \times \square) + 4 = \triangle$
   Pattern: $\frac{2}{3}$
   Slope = ___

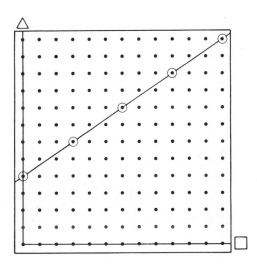

4. Make a conjecture about the relationship between open sentences like those in problem 3 and the slope of the graph of their truth set.

straight line = □ *number × this is slope.*

5. Graph several members of the truth set of the following open sentences, starting with a peg in the vertical column of pegs passing through the origin.

a. □ + 3 = △

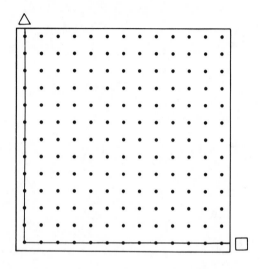

b. □ + 7 = △

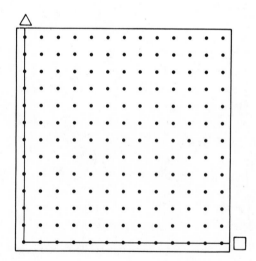

c. $(^-4 \times \square) + 11 = \triangle$

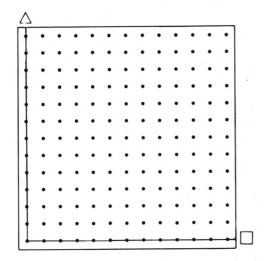

d. $(1/3 \times \square) + 6 = \triangle$

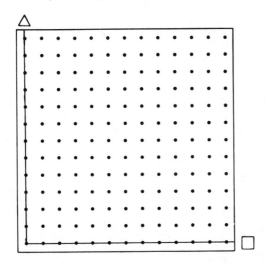

6. Make a conjecture about the relationship between open sentences like those of problem 5 and the location of the peg that is in the column passing through the origin.

plane of solutions

b = intercept

7. Based on your conjecture about the relationships between open sentences and their graphs, write the open sentence for each truth set.

a.

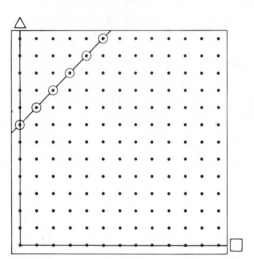

$x + 7 = y$

Sentence:

b.

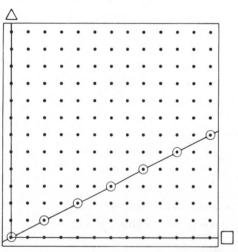

$\frac{1}{2}x = y$

Sentence:

c.

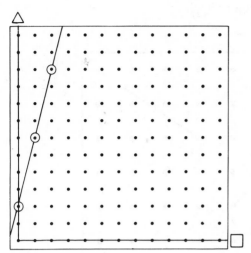

$4x + 2 = y$

Sentence:

d.

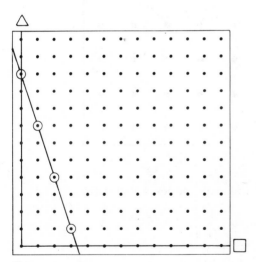

$-3x + 10 = y$

Sentence:

3. Slopes of segments

1. Find the slope of these line segments in Figure 4.

$\overline{EF}$ _____ $\overline{GH}$ _____ $\overline{IJ}$ _____ $\overline{KL}$ _____

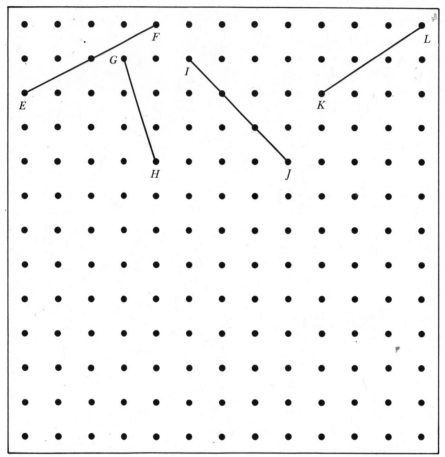

Figure 4

2. In Figure 4, sketch line segments with the following slopes. Label them clearly.

a. 2   b. $\dfrac{3}{5}$

c. $-\dfrac{1}{3}$ d. $\dfrac{1}{3}$

e. $-\dfrac{1}{2}$ f. $\dfrac{3}{2}$

3. Continue the sequence of line segments shown in Figure 5 on a geoboard.

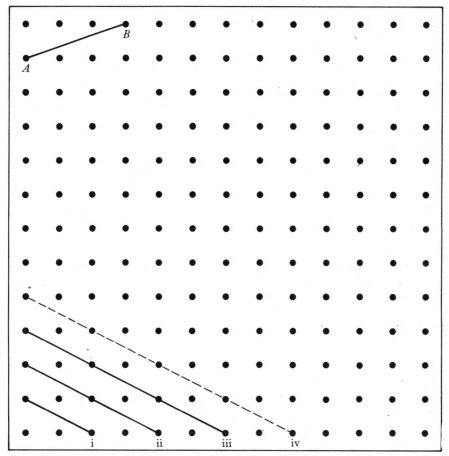

Figure 5

Find the slope of the first six line segments in the sequence.

i. $-\frac{1}{2}$ ii. _____   iii. _____   iv. _____   v. _____   vi. _____

4. The line segments in the sequence are *parallel*. Make a conjecture about the slopes of parallel lines.

5. Starting with line segment $\overline{AB}$, sketch a sequence of parallel line segments on Figure 5.

6. Continue the sequence of line segments in Figure 6(a) on the geoboard.

Find the slope of the first six line segments in the sequence.

i. $\frac{1}{3}$ ii. $-\frac{3}{1}$ iii. $\frac{1}{3}$ iv. $-\frac{3}{1}$ v. _____   vi. _____

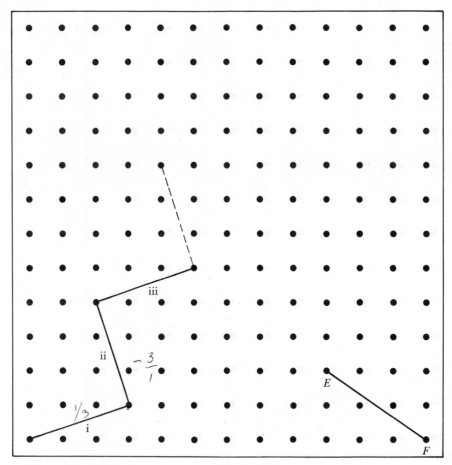

Figure 6(a)

7. The line segments in the sequence are *perpendicular*. Make a conjecture about slopes of perpendicular lines.

$$\frac{3}{1} \times -\frac{1}{3} = -1$$

8. Beginning with line segment $\overline{EF}$, sketch a sequence of perpendicular lines on Figure 6(a).

9. Find the sequence of the slopes for the segments in Figure 6(b).

    a. Slope of $\overline{OA_1}$ = _____ $\frac{1}{4}$

    b. Slope of $\overline{OA_2}$ = _____ $5/6$

c. Slope of $\overline{OA_3}$ = $\frac{2}{3}$

d. Slope of $\overline{OA_4}$ = $\frac{1}{2}$

e. Slope of $\overline{OA_5}$ = $\frac{1}{3}$

f. Slope of $\overline{OA_6}$ = $\frac{1}{6}$

g. Slope of $\overline{OA_7}$ = 0

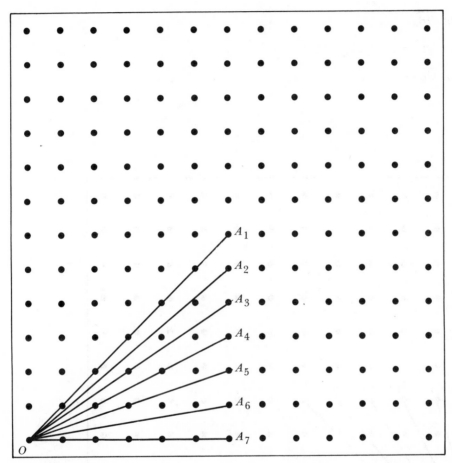

Figure 6(b)

10. Make a conjecture about segments with slopes of 0. Write an open sentence with a truth set represented by a horizontal line.

11. Find the sequence of slopes for the segments in Figure 6(c).

    a. Slope of $\overline{OB_1}$ = ___1___

    b. Slope of $\overline{OB_2}$ = ___4/5___

    c. Slope of $\overline{OB_3}$ = ___3/2___

    d. Slope of $\overline{OB_4}$ = ___2___

    e. Slope of $\overline{OB_5}$ = ___3___

    f. Slope of $\overline{OB_6}$ = ___6___

    g. Slope of $\overline{OB_7}$ = ___no slope___

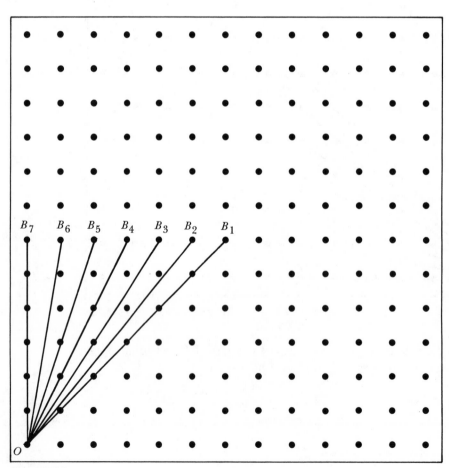

Figure 6(c)

12. Make a decision about the slope of $\overline{OB_7}$ based on the previous discussion about dividing by 0. According to the sequence of slopes, what would you expect for a slope for $\overline{OB_7}$?

13. Imagine a gigantic geoboard with an array of pegs that is 1000×1000. Let the lower left corner peg be the origin (O). Consider the segment OC_1 where C_1 is the peg at $(999, 999)$. The slope of this segment would be $999 \div 999$, or 1. As we move horizontally from C_1 to the left, the slopes of the corresponding segments would be $999 \div 998$, $999 \div 997$, $999 \div 996$, etc. What would the slopes of these segments be when they got close to the vertical line through the origin?

999

14. Write an open sentence whose truth set is a vertical line.

$\square = 1$

4. Intersecting graphs

1. The truth sets for $\square + \triangle = 6$ and $2 \times \square = \triangle$ are graphed on the array in Figure 7. Which replacement is in the truth sets for both sentences?

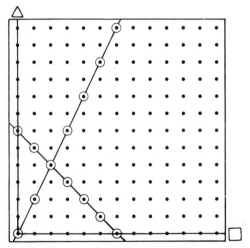

$(2, 4)$

Figure 7

2. Using the graphs, find the truth set for each of the following pairs of number sentences. Determine which replacement is in the truth sets of both sentences.

EXAMPLE $(3 \times \square) + 1 = \triangle$
$\triangle + \square = 9$
The replacements (2, 7) are in the truth set for both equations.

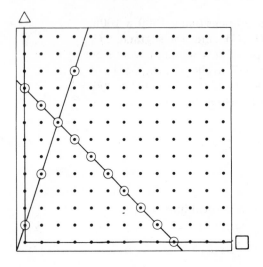

a. $\square + 1 = \triangle$
$(3 \times \square) - 3 = \triangle$

2, 3

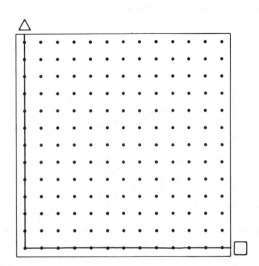

b. $(4 \times \square) + 5 = \triangle$
   $(^-2 \times \square) + 5 = \triangle$

0, 5

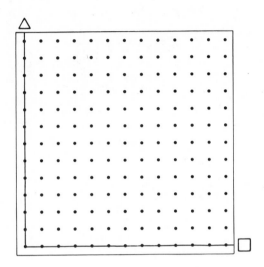

c. $\square + 6 = \triangle$
   $(3 \times \square) + 2 = \triangle$

2, 8

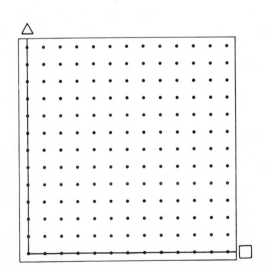

d. $(2 \times \square) - 5 = \triangle$
   $(^-2 \times \square) + 7 = \triangle$

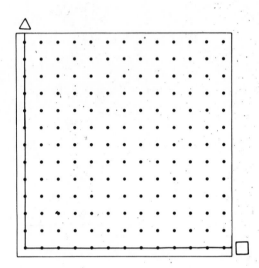

3, 1

e. $(3 \times \square) - 2 = \triangle$
   $(^-2 \times \square) + 8 = \triangle$

2, 4

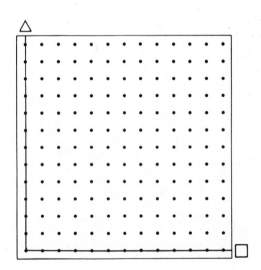

3. Graph the truth sets for the following pairs of number sentences. Identify the replacements that are in the truth sets of both open sentences.

EXAMPLE $\triangle + \square > 7$
$(1/2 \times \square) + 3 > \triangle$
The double shaded region in Figure 8 is a graph of the replacements that are in the truth sets of both sentences.

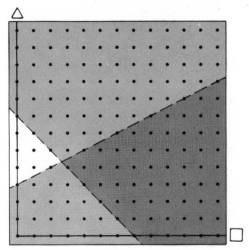

Figure 8

a. $\square < \triangle$
$\triangle + (2 \times \square) < 6$

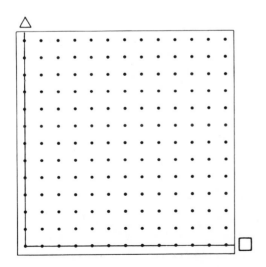

b. $(1/2 \times \square) + 6 < \triangle$
$(3 \times \square) + 1 < \triangle$

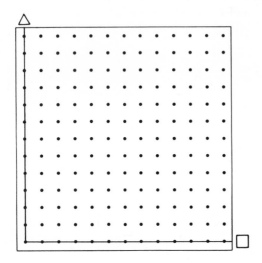

c. $\triangle + \square < 9$
$\square + 3 < \triangle$

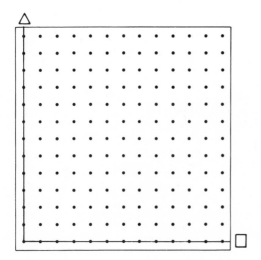

5. *Graphs of curves*

We can extend the scope of graphing to include pairs that have negative numbers by moving the origin to the center of the array as shown in Figure 9. The sign of the numbers then distinguishes up from down and right from left. A negative number indicates the number of units left of or down from this central origin.

EXAMPLE   ($^-$3, 2) is at *A*

($^-$5, $^-$1) is at *B*

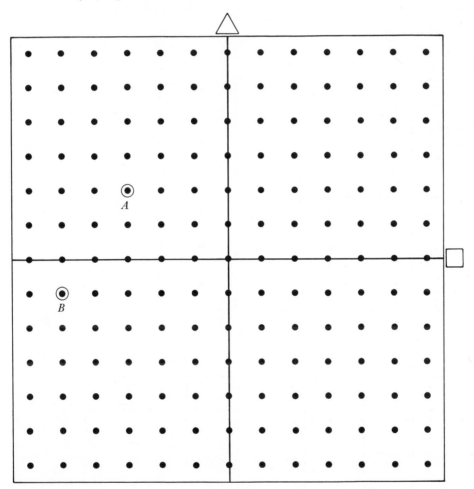

Figure 9

The *row* of dots through the origin is called the *horizontal axis* and is often named by the symbol used to denote the left–right units. In this case it would be the □ axis. The *column* of dots through the origin is called the *vertical axis* and is likewise named the △ axis.

1. Graph the following pairs in the array of Figure 9:

   a. $(4, {}^-3)$

   b. $({}^-1, 4)$

   c. $({}^-5, {}^-3)$

   d. $(3, 2)$

   e. $(6, {}^-4)$

2. Find several members of the truth sets for the following sentences and graph them.

EXAMPLE $(\square \times \square) = \triangle$

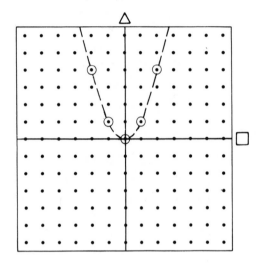

$\square$	$\triangle$
0	0
1	1
${}^-1$	1
2	4
${}^-2$	4

   a. $(\square \times \square) - (6 \times \square) + 8 = \triangle$

$\square$	$\triangle$

b. $(\square \times \square) - (8 \times \square) + 15 = \triangle$

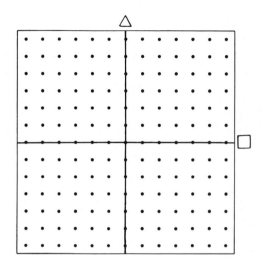 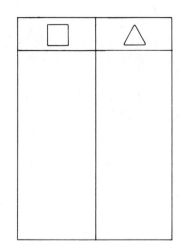

c. $(\square \times \square) + (2 \times \square) = \triangle$

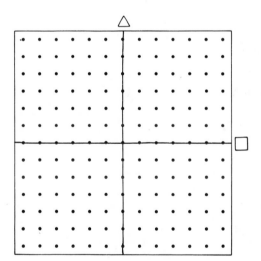 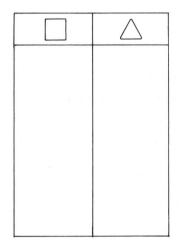

3. The graphs called for in problem 2 are called *parabolas*. Figure 10 is the graph of the truth set for:

$(\square \times \square) - (3 \times \square) + 2 = \triangle$

   a. Determine where the graph intersects the horizontal axis.

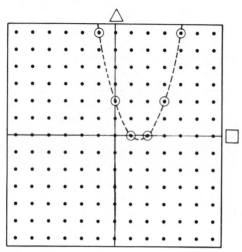

Figure 10

b. Place the values just determined for □ in the sentence

(□ × □) − (3 × □) + 2 = △

What is the truth set for this sentence?

4. Using the graphs, find the truth set for the following:

a. (□ × □) − (7 × □) + 10 > △

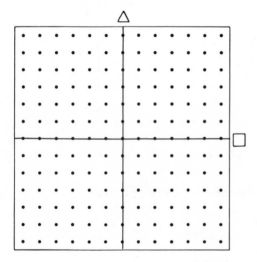

b. $(\square \times \square) - (5 \times \square) + 6 < \triangle$

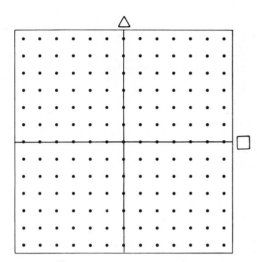

5. Graph the truth set for the following equations as in Figure 11.

EXAMPLE $(\square \times \square) + (\triangle \times \triangle) = 25$

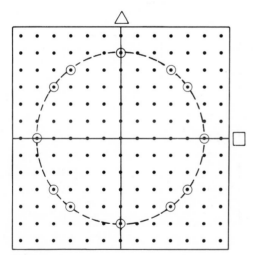

Figure 11

$\square$	$\triangle$
0	5
0	⁻5
5	0
⁻5	0
3	4
3	⁻4
⁻3	4
⁻3	⁻4
4	3
4	⁻3
⁻4	3
⁻4	⁻3

a. $(\square \times \square) + (\triangle \times \triangle) = 13$

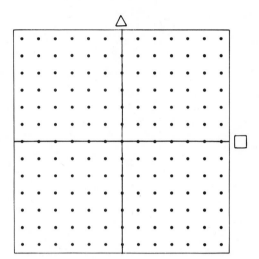 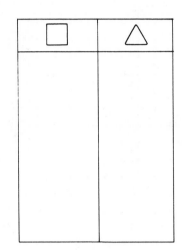

b. $(\square \times \square) + (\triangle \times \triangle) = 20$

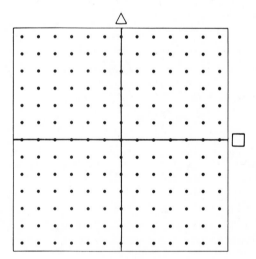

c. $\left[(\square - 1) \times (\square - 1)\right] + (\triangle \times \triangle) = 25$

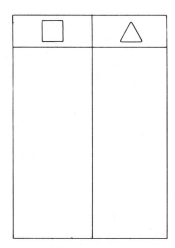

Summary

Using simple equations and a geoboard is an efficient way to develop a basis for understanding equations and graphs. Note the difference in the means we have used to deal with equations. First we generated sentences from a number balance. The balance itself only *indicated* whether we had an equation or an inequality. The numbers on the balance and the number of clips at each position had a special mathematical relationship that permitted us to calculate beforehand whether the arm would balance.

Second we used the geoboard as a means of physically modeling the relationships among the unknowns of open sentences. This is a much more complex situation than showing whether certain combinations of paper clips and numbers balanced. The graph allowed us to see the pattern clearly and to predict how it would continue.

Third we used a table to list pairs of numbers that made the sentences true. While this was helpful, it did not give us the visual picture we got from a graph on a coordinate system.

At this point we suggest you discuss the problems you worked on in Chapter 1. Perhaps now you can gather, record, and interpret information on those activities in a more thorough and confident way. Gathering data from the balance was really quite simple, but learning to deal with practical situations requires much more skill and understanding. The only effective way to become competent in this area is to practice the art and skill of data gathering, recording, and interpreting in real situations.

References

Magnuson, Russell. "Signed Numbers." *The Arithmetic Teacher,* 13 (November 1966), 573–575.

Sganga, Francis T. "A Bee on a Point, a Line, and a Plane." *The Arithmetic Teacher,* 13 (November 1966), 549–552.

12

The language of sets

Sets—collections of objects, including "objects" of thought, events, and the like—are something we deal with daily. School children work with sets to gain insight into such basic concepts as "oneness," "twoness," and "three-ness." High school students use the language and symbolism of sets in their study of algebra and geometry, and we all show awareness of sets in the way we order our lives, from sorting silverware to agreeing on the meaning to be given sentences that contain the logical operators "and," "or," and "not." Because sets pervade our lives, we should know something about them. The objective of this chapter is to provide you with that "something."

1. Classification

The objects that make up a set are typically grouped in terms of their *sameness* with respect to one or more distinctive features. A set of tall, thin, blue-eyed boys, for example, would consist of boys who are the same in these respects, regardless of their differences otherwise. The defining characteristics of this set—tallness, thinness, and blue-eyedness—are referred to as *attributes,* and the selection of objects with respect to specific attributes is called *classification.* By completing the following exercises with the attribute shapes (the 32 die-cut geometric figures in the insert at the end of the book), you should gain insight into classification and the way sets relate to one another.

1. Sort the attribute shapes into 2 distinct sets. Give the attribute(s) the figures in each set have in common.

Repeat for 4 distinct sets, 8 distinct sets, and 16 distinct sets.

2. Sort the attribute shapes into sets so that the figures in each set are alike in size and color. How many sets are there? _____ How many figures are in each set? _____ The figures in each set are alike in that they are either all _____, all _____, all _____, all _____, all _____, all _____, all _____, or all _____.

3. Sort the attribute shapes into sets so that the figures in each set are alike in size, shape, and color. How many sets are there? _____ How many figures are in each set? _____ Describe the effect of increasing the number of attributes on the number of objects that make up a set.

4. Complete the following pattern for the large figures. The letters *R, G, Y,* and *B* represent the colors red, green, yellow, and blue, respectively.

R			R
	G		
		Y	
			B

What attributes do the figures in each row of the array have in common?

What attributes do the figures in each column of the array have in common?

5. Using all the attribute shapes, lay out a figure-8 in which adjacent figures are alike in only two ways. Which part of the figure-8 was the hardest to construct? Why? Could a figure-8 be constructed if adjacent figures were to be alike in only one way? In no ways?

6. Two-sameness trains and crossings are arrangements of attribute shapes wherein adjacent figures are the same in exactly two ways. One-sameness and no-sameness trains and crossings are defined similarly. Complete the trains and crossings suggested in the following diagrams:

Two-sameness train

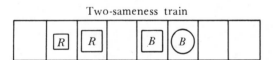

One-sameness train

R

No-sameness train

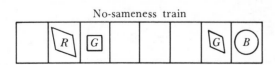

One-sameness crossing

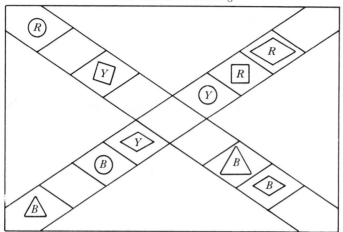

Two-sameness crossing

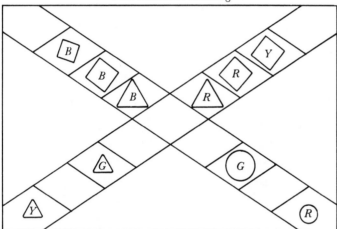

No-sameness crossing

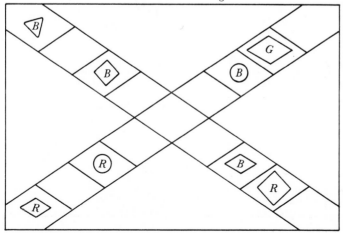

Compare your trains and crossings with those of a neighbor. Are they the same? Why or why not?

7. Select an opponent for playing "Close Out." The first player puts an attribute shape in one of the 16 spaces of Figure 1. The second player puts another attribute shape in any of the remaining spaces in accordance with the following rule: If a figure lies directly above or below or to the immediate right or left of another figure, it must be the same as that figure in

Figure 1

exactly two ways. The game continues in this manner until either the playing board is covered or all possible plays have been made. The winner is the person making the last play.

Play the game again in accordance with the rule that neighboring attribute shapes must be the same in exactly one way. Make a conjecture about a winning strategy for this game.

2. Construction of sets

In this section the attribute shapes are used to illustrate the concepts *subset, intersection, union,* and *complement.* A glimmer of understanding of these terms is sufficient at this point. Their exact meaning will be made clear in the next section.

1. A circle dart game with a score of 12 is illustrated in Figure 2. The five "darts" in the outside ring are worth a total of 5 points since each one satisfies one criterion, that of being circular. The two "darts" in the middle ring are worth a total of 4 points since each one satisfies two criteria, those of being circular and small. The "dart" in the bull's eye counts 3 points since it satisfies three criteria, those of being circular, small, and red.

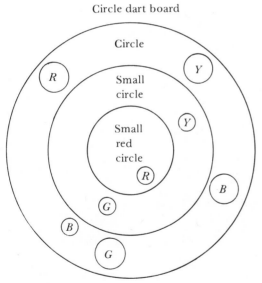

Figure 2

How should the "darts" in Figure 2 be placed for a maximum score of 13?

For a minimum score of 8?

How could the circle dart board in Figure 2 be relabeled so that the most that could be scored would be 11?

Label the "red" dart board and the "small" dart board that follow, and give the maximum and minimum scores possible for each. How are the sets that make up each of the dart boards related?

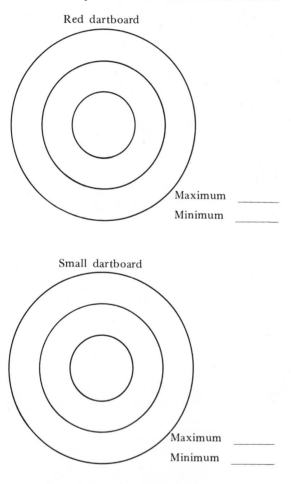

Red dartboard

Maximum _____
Minimum _____

Small dartboard

Maximum _____
Minimum _____

2. For the following circles, would it be possible to stack all the triangular figures in the circle on the left and all the square figures in the circle on the right? Why or why not?

Would it be possible to stack all the green figures in the circle on the left and all the square figures in the circle on the right? Why or why not?

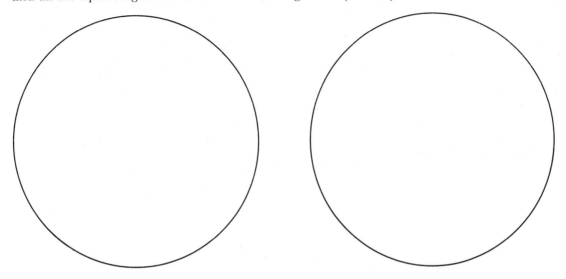

3. Suppose we were to overlap the two circles in problem 2 as indicated in the following diagram. Would it then be possible to stack all the green figures in the circle on the left and all the square figures in the circle on the right? Where would you stack the green, square figures?

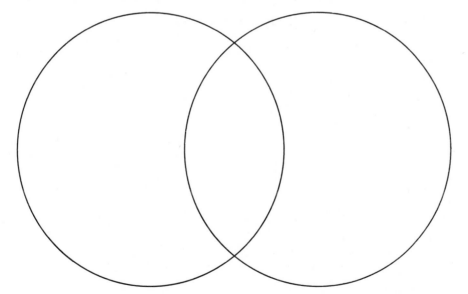

How does the set of green, square figures relate to the set of green figures and the set of square figures?

4. The letter U in Figure 3 refers to the *universal* set, the set of everything under examination. Typically, U contains only those elements that encompass all the particulars being considered. For the elements under examination in Figure 3, U is the set of all the square figures and all the yellow figures since squareness and yellowness are the only particulars being considered. It is *not* the set of all 32 attribute shapes since that set contains a lot of figures exhibiting characteristics other than squareness and yellowness.

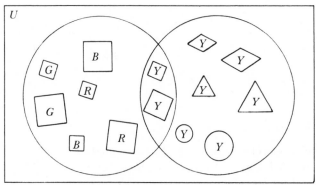

Figure 3

Describe U for the sets under consideration in the following diagrams:

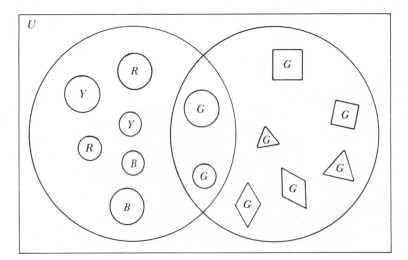

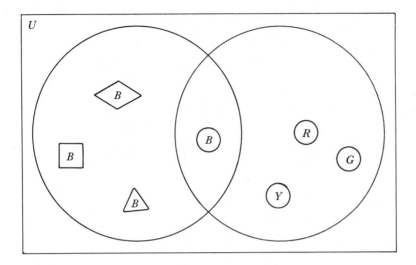

5. A three-way sort is shown in Figure 4. Describe a three-way sort of your own in the three overlapping circles provided.

Three-way sort

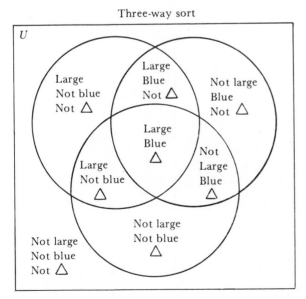

Figure 4

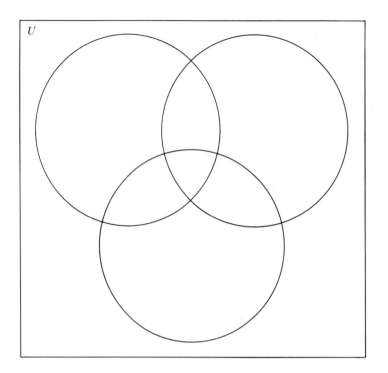

6. Supply the missing contents.

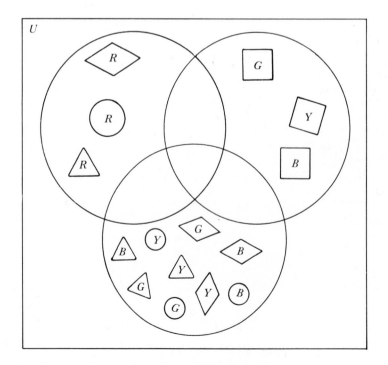

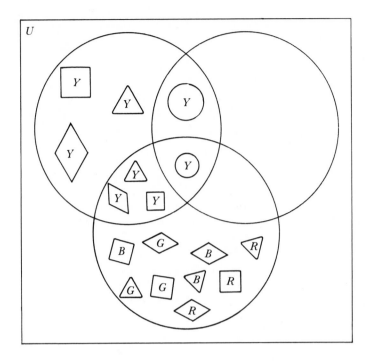

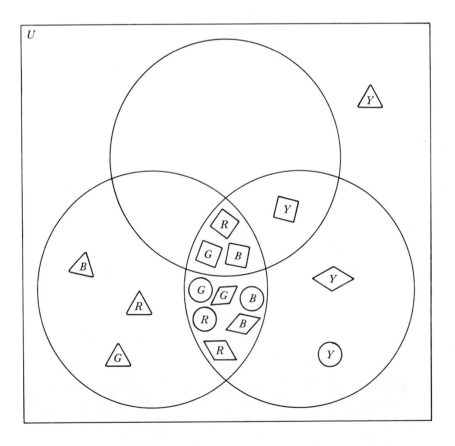

How does the set in the middle of each triple of overlapping circles relate to the sets that make up the triple?

3. Language and notation of sets

There are two ways of denoting a set. One way is to list the objects that make up the set. The other way is to give a rule describing the objects that make up the set. In either case, what is written is enclosed in braces. To illustrate, in list form the set of small figures would be denoted as {small blue circle, small green circle, small red circle, small yellow circle, small blue parallelogram, small green parallelogram, small red parallelogram, small yellow parallelogram, small blue square, small green square, small red square, small yellow square, small blue triangle, small green triangle, small red triangle, small yellow triangle}. In rule form, the set would be denoted as {small figures}. The set made up of nothing is denoted as { }, braces enclosing nothing, or as $\emptyset$, a symbol similar to 0 in its appearance and meaning. The symbol $\emptyset \neq 0$; however, since $\emptyset$ is a set and 0 is a number. Also $\{\emptyset\} \neq \emptyset$ since $\{\emptyset\}$ is made up of something, namely, $\emptyset$.

1. Denote the set of triangular figures by listing its elements.

{L, red A,   S red A, ...}

2. Denote the set of red figures by giving a rule.

{Red figures}

3. Denote the set of green circular figures by listing the elements or by giving a rule.

4. Which method for denoting a set do you prefer, the list method or the rule method? Why?

The objects that make up a set are called the *elements* of the set. The elements of the set of small circular figures, for example, would be the small blue circle, the small red circle, the small yellow circle, and the small green circle.

To denote that something is an element of a set, we use the symbol ϵ, the lowercase epsilon of the Greek alphabet. To denote that the small blue circle is an element of the set of small circular figures, for example, we write "small blue circle ϵ {small circular figures}."

Whenever all the elements of one of two sets are the same as some or all the elements of the other set, the one set is said to be a *subset* of the other set. The set of small square figures is a subset of the set of squares, for example, in that every small square in one is the same as one of the squares in the other.

To denote that a set is a subset of another set, we use the symbol $\subseteq$. To denote that the set of small square figures is a subset of the set of squares, for example, we write "{small square figures} $\subseteq$ {square figures}."

The primary distinction to be made between ϵ and $\subseteq$ is this: The symbol ϵ relates an element to a set; $\subseteq$ relates a set to a set.

5. Insert either ϵ, $\notin$ (is not an element of), $\subseteq$ or $\nsubseteq$ (is not a subset of) to make each of the following a *true* statement:

a.   small red square ϵ {small figures}

b.   {large square figures} $\subseteq$ {square figures}

c. $\varnothing \subseteq$ {red figures}

d. $\varnothing \subseteq$ {square figures}

e.   {blue circular figures} $\subseteq$ {blue figures}

f.   {small yellow figures} $\subseteq$ {yellow figures}

g.   {triangular figures} $\subseteq$ {triangular figures}

h.   square ϵ {square figures}

i.   red circle ϵ {small circular figures}   *or* $\notin$

j.   green circle ϵ {circular figures}   *or* $\notin$

k.   square ϵ {blue square figures}   *or* $\notin$

l.   {triangular figures} $\nsubseteq$ {large triangular figures}

m.   {red figures} $\nsubseteq$ {red circular figures}

n.   red small parallelogram ϵ {red figures} $\notin$

o.   red triangle ϵ {red figures}

p.   {green square figures} $\subseteq$ {square figures}

The principal operations on sets are intersection, union, and complement. The *intersection* of two sets is the set of elements belonging to *both* of the sets. For example, the intersection of the set of yellow figures and the set of square figures, denoted {yellow figures} $\cap$ {square figures}, is the set of figures that are yellow *and* square: {small yellow square, large yellow square}. Whenever none of the elements of one of the two sets belong to the other set, the intersection of the sets is $\varnothing$, and the sets are said to be *disjoint*. For example, the set of yellow figures and the set of red figures are disjoint, because none of the yellow figures belong to the set of red figures.

The *union* of two sets is the set of elements belonging to *either* of the sets. For example, the union of the set of yellow figures and the set of square figures, denoted {yellow figures} ∪ {square figures}, is the set of figures that are yellow *or* square: {small yellow circle, large yellow circle, small yellow triangle, large yellow triangle, small yellow parallelogram, large yellow parallelogram, small yellow square, large yellow square, small blue square, large blue square, small red square, large red square, small green square, large green square}.

The *complement* of a set is the set of elements belonging to a universal set but *not* to the set itself. For example, the complement of the set of small circular figures, denoted {small circular figures}′, for the universe of circular figures is the set of figures that are circular but not small: {large yellow circle, large blue circle, large red circle, large green circle}.

6. Perform the indicated operations:

   a. {small figures} ∩ {circular figures}

   b. {blue figures} ∩ {triangular figures}

   c. {square figures} ∩ {square figures}

   d. {small red figures} ∩ {small green figures}

   e. ∅ ∩ {large square figures}

   f. {small figures} ∪ {circular figures}

   g. {blue figures} ∪ {triangular figures}

   h. {square figures} ∪ {square figures}

   i. {small red figures} ∪ {small green figures}

   j. ∅ ∪ {large square figures}

   k. {yellow circular figures}′ for U = circular figures

l. {yellow circular figures}′ for U = yellow figures

m. {small figures}′ for U = the set of all 32 attribute shapes

n. U' for U = the set of all 32 attribute shapes

o. $\emptyset'$ for U = the set of all 32 attribute shapes

4. *Properties of sets*

1. Complete the intersection and union charts for: $\emptyset$, $R = \{red\ figures\}$, $X = \{red\ circular\ figures\}$, $Y = \{red\ or\ circular\ figures\}$, $C = \{circular\ figures\}$, and U = the set of all 32 attribute shapes.

∩	∅	R	X	C	U	Y
∅	∅	∅	∅	∅	∅	∅
R	∅	R	X	X	R	R
X	∅	X	X	X	X	X
C	∅	R	X	C	Y	Y
U	∅	X	X	C	C	C
Y	∅	R	X	X	U	Y

∪	∅	R	X	C	U	Y
∅	∅	R	X	C	U	Y
R	R	R	R	Y	U	Y
X	X	R	X	C	U	Y
C	C	Y	C	C	U	Y
U	U	U	U	U	U	U
Y	Y	Y	Y	Y	U	Y

Problems 2–10 refer to the intersection and union charts you completed in problem 1.

2. Is the set $\{\varnothing, R, X, Y, C, U\}$ closed under ∩? Under ∪? How do you know?

yes　　yes

3. Is ∩ associative? Commutative? Justify your assertions.

y　　y

4. Is ∪ associative? Commutative? Justify your assertions.

y　　y

5. Is there an identity element for ∩? For ∪?

∪　　∅

6. Are there inverses for ∩? For ∪?

only for ∪　only for ∅

7. Does the set $\{\varnothing, R, X, Y, C, U\}$ form a group under ∩? Under ∪? Justify your assertions.

no　　no

8. Does ∩ distribute over ∪? Justify your assertion.

yes

9. Does ∪ distribute over ∩? Justify your assertion.

yes

10. Does the set $\{\varnothing, R, X, Y, C, U\}$ form a field under ∩ and ∪? Justify your assertion.

no　　inverse

5. *Meaningful counting*

Sets are useful for introducing much of the language and symbolism of mathematics, but the most important use of sets is in giving meaning to *counting*. Consider the set of square figures, for example, and two similar-looking statements that describe this set: (1) There are only squares in the set, and (2) There are eight squares in the set. The first statement could be verified by looking at each of the elements of the set to see if it were a square. The second statement could not be verified in this manner, however, since to attempt to do so would be to look at each of the elements of the set to see if it were an "eight square" or not. To verify the second statement, the figures that make up the set would have to be counted because the set itself, rather than the squares that make up the set, has the property of "eightness."

To investigate the numerical properties of sets, we introduce the concept of *equivalence*. Two sets are said to be "equivalent" whenever they contain the same number of elements. For example, the set of square figures and the set of circular figures are equivalent since each set contains eight elements; that is, each set has the property of "eightness."

To indicate that the set of square figures and the set of circular figures have the property of "eightness," we say that their cardinal number is 8. The *cardinal number* of a set is the number of elements that make up the set.

1. Determine the cardinal number of each of the following sets:

   a. {small figures}   *16*

   b. {yellow figures}   *8*

   c. {red figures}   *8*

   d. {large blue figures}   *4*

   e. {large circular figures}   *4*

   f. {small green figures}   *4*

   g. {circular figures} ∩ {green figures}   *2*

   h. {square figures} ∪ {triangular figures}   *16*

   i. {large yellow square figures}   *1*

   j. {small green circular figures}   *1*

   k. {circular figures} ∪ {red figures}   *14*

   l. {small figures} ∩ {triangular figures}   *4*

   m. {small figures} ∪ {circular figures}   *20*

   n. {square figures}   *8*

   o. {large yellow triangular figures}   *1*

   p. {small red figures}   *4*

   q. ∅   *0*

   r. {1, 2}   *2*

s. $\{1, 2, 3, 4\}$ 4

t. $\{1, 2, 3, \ldots, 14, 15, 16\}$ 16

u. $\{1, 2, 3, 4, 5, 6, 7, 8\}$ 8

v. $\{8, 7, 6, 5, 4, 3, 2, 1\}$ 8

w. $\{0, 1, 2, 3, 4, 5, 6, 7\}$ 8

x. $\{8\}$ 1

y. $\{0\}$ 1

z. $\{\ \}$ 0

2. Which of the sets in problem 1 are equivalent?

3. Under what conditions does the cardinal number of the union of two sets equal the sum of the cardinal numbers of the sets?

Disjoint sets

4. Under what conditions does the cardinal number of the union of two sets equal the sum of the cardinal numbers of the sets minus the cardinal number of the intersection of the sets?

6. *Sets and logic*

Logic is the science of the use of language in argument and persuasion. It deals with those parts of speech that require clarification for effective communication. In particular, it deals with the logical operators *and, or,* and *not.*

The relationship between logic and set theory is that expressions containing "and," "or," and "not" can be illustrated with sets and the operations of intersection, union, and complement, respectively. For example, the expression "red figures and square figures" could be viewed as the intersection of the set of red figures and the set of square figures since the intersection of two sets is the set of elements possessing the attributes common to the elements of the first set *and* the second set. Similarly, an expression containing the word "or" could be viewed as the union of two sets since the union of two sets is the set of elements possessing the attributes of the elements of the first set *or* those of the second set. And an expression containing the word "not" could be viewed as the complement of a set since the complement of a set is the set of elements that do *not* possess the attributes of the elements of the set.

The advantage of illustrating expressions with sets is that in set form they can be simplified. Take the expression "red figures and square figures," for example. In set form it would appear as the intersection of the set of red figures and the set of square figures, which would be the set of red square figures, which would correspond to the expression "red square figures," a clarification of the original expression.

1. Translate the following expressions into set language and simplify whenever possible for U = the set of all 32 attribute shapes.

   a. blue figures and triangular figures

   b. small figures or red figures

   c. not large figures

   d. not not large figures

   e. blue figures and not square figures

   f. not small figures and not green figures

   g. neither small figures nor green figures

   h. red figures or not square figures

   i. not red figures or not square figures

2. Translate the following identities into logical terms for sets A and B, any subsets of the set of all 32 attribute shapes.

   a. $A \cup A' = U$

   b. $A \cap A' = \varnothing$

   c. $(A')' = A$

d. $(A \cup B)' = A' \cap B'$

e. $(A \cap B)' = A' \cup B'$

Summary

Having completed this chapter, it is important to understand the perspective from which mathematicians invariably include something on sets in any development of mathematics. Sets provide the point of departure for thinking about the logical development of all mathematics. They are cornerstones of each of the major areas of mathematics—arithmetic, algebra, and geometry—and the language and symbolism associated with sets facilitate communication in these areas. In arithmetic, for example, addition of whole numbers can be viewed in terms of the union of two disjoint sets. In algebra the solution of one equation in one unknown can be regarded as the set of all real numbers satisfying the condition on the unknown. And in geometry the intersection of two lines can be thought of as the point which the two sets of points that make up the two lines have in common.

The primary importance of sets in mathematics is that they can be a beginning point in the development of thinking skills. Since we consider mathematics to be an efficient, useful way of thinking, it is only appropriate that we study sets. A better understanding of sets will provide a basis for helping others learn about arithmetic, algebra, and geometry.

References

Elementary Science Study. *Attribute Games and Problems* (Teacher's Guide), Webster Division, McGraw-Hill Book Company, New York, 1966.

Dienes, Z., and E. W. Golding. *Modern Mathematics for Young Children.* Herder and Herder, New York, 1970.

Dienes, Z., and E. W. Golding. *Sets, Numbers and Powers.* Herder and Herder, New York, 1966.

13

Probability and statistics

This chapter deals with the mathematics of chance—the study of probability. A major objective is to acquaint you with ways of determining the likelihood that an event will take place. *Likelihood* means that fraction of time the event is expected to take place.

1. A friendly game of chance

A gambling house advertises a friendly game of chance: The house tosses a coin. If it comes up heads, you pay the house $1.00, and the house tosses the coin again. If it comes up tails, you get to toss the coin. If it comes up heads, the house pays you $2.00, and the coin returns to the house. If it comes up tails, you pay the house $1.00, and the coin returns to the house.

Do you think this game is fair? To find out, play the game with someone from your class. One of you be the house, the other, the gambler. Take 20 turns and record your results in Table 1.

  a. Who won the most money? How much was won?

  b. Who won the most money for your class as a whole?

*Table 1   Results of playing a friendly
game of chance*

Turn	House's coin	Gambler's coin	Winner
1			
2			
3			
4			
5			
6			
7			
8			
9			
10			
11			
12			
13			
14			
15			
16			
17			
18			
19			
20			

c. Was the game fair?

d. How many dollars do you think the gambler should have received each time he or she won?

e. The house argues that there are three possibilities—1 head (house wins), 2 tails (house wins), and a tail and a head (gambler wins)—and that since only two of these favor the house, it can expect to win only $\frac{2}{3}$ of the time. The gamblers say that the house can expect to win $\frac{3}{4}$ of the time since 1 head will occur $\frac{1}{2}$ of the time and 2 tails will occur $\frac{1}{4}$ of the time. Which argument do your results support?

Which argument do you think is correct?

2. *Likely versus unlikely outcomes*

An *outcome* is any of the ways in which an activity can end. For example, the outcomes for tossing a coin once are a head and a tail. A set of outcomes is called an *event*. The events for tossing a coin once would therefore be {head} and {tail}. Central to determining the likelihood of an event is whether the outcomes that make up the event can be expected to occur with equal frequency. As the following exercises illustrate, the outcomes of some activities can be expected to occur with equal frequency, and the outcomes for other activities cannot.

Experiment 1

If you were to observe the last digit of 100 telephone numbers from any page of a telephone directory, which digit do you think would occur most often? To find out, perform the activity and record your findings in Table 2.

Table 2 *Results of observing the last digit of 100 telephone numbers*

Outcome	Tally			Frequency		
0	15	11	13	15	7	13
1	15	17	6	5	14	14
2	11	7	8	10	9	8
3	10	11	10	9	9	8
4	11	11	5	10	15	13
5	9	11	11	13	8	15
6	9	14	11	12	13	9
7	9	6	9	10	9	3
8	6	7	13	9	10	9
9	5	9	11	7	6	8
				Total = 100		

a. How many times did the digit you chose occur?

b. About what fraction of the time did it occur?

c. Would you select this digit to occur most often as the last digit for the next 100 telephone numbers?

d. Make a conjecture about the likelihood of one digit occurring as the last digit in a telephone number compared to another digit.

Illustrate your results for Table 2 with a bar graph in the space provided.

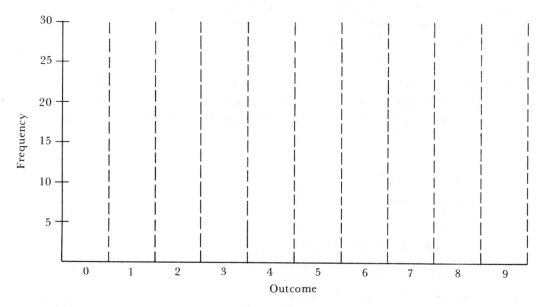

Experiment 2

Suppose you had observed the *first* digit of the 100 telephone numbers you used for experiment 1. Would that have altered your results for that experiment? To find out, perform the activity and record your findings in Table 3.

a. Did the results of this experiment differ much from those of experiment 1?

Table 3   Results of observing the first digit of 100 telephone numbers

Outcome	Tally	Frequency
0		
1		
2		
3		
4		
5		
6		
7		
8		
9		
		Total = 100

b. Make a conjecture about the likelihood of one digit occurring as the first digit in a telephone number compared to another digit.

Illustrate your results for Table 3 with a bar graph in the space provided.

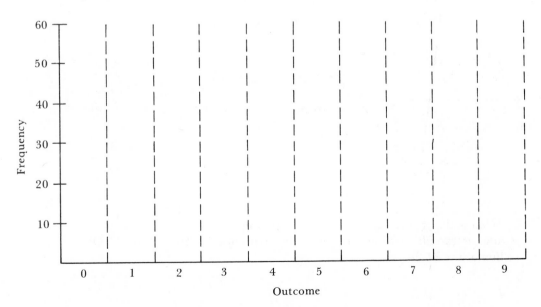

3. *Theoretical probability*

The *probability* of an event is a number that reflects the likelihood of the event's occurring. For events made up of outcomes equally likely to occur, the probability is simply the ratio of the number of outcomes that make up the event to the number of possible outcomes. The probability of rolling a 3 with a single die is therefore $\frac{1}{6}$ since there is exactly 1 way to get a 3 among 6 equally likely outcomes, namely 1, 2, 3, 4, 5, and 6. Similarly, the probability of rolling a sum of 8 with a pair of dice is $\frac{5}{36}$ since there are exactly 5 ways to get a sum of 8 from 36 equally likely outcomes as illustrated in the following *array*.

(1, 1)	(1, 2)	(1, 3)	(1, 4)	(1, 5)	(1, 6)
(2, 1)	(2, 2)	(2, 3)	(2, 4)	(2, 5)	(2, 6)
(3, 1)	(3, 2)	(3, 3)	(3, 4)	(3, 5)	(3, 6)
(4, 1)	(4, 2)	(4, 3)	(4, 4)	(4, 5)	(4, 6)
(5, 1)	(5, 2)	(5, 3)	(5, 4)	(5, 5)	(5, 6)
(6, 1)	(6, 2)	(6, 3)	(6, 4)	(6, 5)	(6, 6)

Since the probabilities $\frac{1}{6}$ and $\frac{5}{36}$ in the foregoing examples were obtained theoretically (that is, without experimentation), they are called *theoretical* probabilities.

1. What is the theoretical probability of rolling a 3 with a regular tetrahedron whose faces have been numbered from 1 to 4? Of rolling a 4?

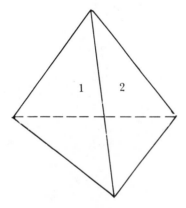

2. What is the theoretical probability of rolling a sum of 5 with a pair of regular tetrahedra whose faces have been numbered from 1 to 4? Of rolling a sum of 6?

3. A hat contains 10 slips of paper numbered from 1 to 10. You are to select a number from the hat blindfolded.

   a. What is the theoretical probability of selecting a number that is divisible by 2?

   b. By 3?

   c. By 5?

4. There are 4 black marbles, 3 white marbles, 2 blue marbles, and 1 red marble in a box. You are to select a marble without looking.

   a. What is the theoretical probability of selecting a black marble?

   b. A red marble?

   c. A marble that is either black or red?

   d. A marble that is not white?

   e. A marble that is either black, white, blue, or red?

   f. A green marble?

5. A spinner is numbered from 1 to 10 as illustrated.

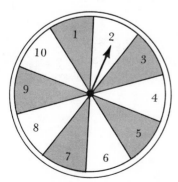

   a. What is the theoretical probability of spinning a 10?

b. A 7 or less?

c. A number greater than 7?

d. An odd number?

6. A standard deck of playing cards contains 52 cards, 13 in each of 4 suits: spades, hearts, diamonds, and clubs.

a. If the deck is shuffled, what is the probability of drawing a 3 of hearts?

b. A 9 of diamonds?

c. A jack of clubs?

d. A 4 of any suit?

e. An ace of any suit?

f. A heart?

g. A spade or club?

4. *Empirical probability*

Not all outcomes of an experiment are equally likely to occur. For example, the outcomes of tossing a thumbtack are "point up" and "point down," and there is no reason to expect that these outcomes will occur with equal frequency.

Point up   Point down

In fact, for most thumbtacks they won't. Therefore, to assign a probability to, say, "point up," it is necessary to toss a thumbtack a number of times (the more, the better) and take as its probability the ratio of the number of "point up" occurrences to the total number of tosses. Since this number would come from experimentation, it would be called an *empirical* probability.

Experiment 3

Shake 10 thumbtacks in a cup and turn the cup face down on your desk. Record the number of thumbtacks landing point up and the number landing point down in Table 4 on page 314. Repeat 25 times.

a. Using the results of your experiment, what is the empirical probability of a thumbtack landing point up? Point down?

b. What is the sum of your probabilities for point up and point down?

c. How do your probabilities for point up and point down compare with some of your classmates' probabilities?

d. If you tossed a thumbtack 1000 times, how many times would you expect it to land point up?

Experiment 4

A paper cup can land in three ways: on its small end, on its large end, or on its side.

Small end   Large end   Side
down   down

Toss a paper cup 50 times and in Table 5 on page 315, record the number of times it lands in the various positions.

a. What did you get for the empirical probability of a cup landing small end down? Large end down? On its side?

b. What is the sum of these probabilities?

c. Make a conjecture about the sum of the probabilities of all possible events of any experiment.

Table 4   Results of tossing 10 thumbtacks 25 times

Turn	Number of thumbtacks landing point up	Number of thumbtacks landing point down
1		
2		
3		
4		
5		
6		
7		
8		
9		
10		
11		
12		
13		
14		
15		
16		
17		
18		
19		
20		
21		
22		
23		
24		
25		
	Total =	Total =

Table 5 Results of tossing a paper cup 50 times

Outcome	Tally	Frequency
Small end down		
Large end down		
Side		
		Total = 50

d. Make a conjecture about the probability of a cup landing small end down.

Experiment 5

Were we to toss a penny repeatedly, we would expect to get just as many heads as tails. What would you expect to get if you *spun* the penny instead? To check your conjecture, spin a penny 50 times and record your results in Table 6.

To spin, hold with one forefinger and snap with the other forefinger.

Table 6 Results of spinning a penny 50 times

Outcome	Tally	Frequency
Head		
Tail		
		Total = 50

a. What did you get for the empirical probability of a spun penny coming up heads? Tails?

b. How do your probabilities compare with some of your classmates' probabilities?

c. Does the age of the penny appear to have anything to do with the results?

d. Does spinning a penny make for a fair game?

Experiment 6

The following is a code left by Captain Kidd. It directs the translator to a buried treasure as related by Edgar Allan Poe in "The Gold Bug."

53‡‡†305))6*;4826)4‡.)4‡) ;806*;48†8¶60))85;1‡(;:‡*8†83(88
5*†;46(;88*96*?;8)*‡(;485) ;5*†2:*‡(;4956*2(5*—4)8¶8*;40692
85);)6†8)4‡‡;1(‡9;48081;8:8‡1;48†85;4)485†528806*81(‡9;48;(8
8;4(‡?34;48)4‡;161;:188;‡?;

A simple strategy for breaking codes is to replace the symbols that occur most often in a code with the letters that tend to occur with the same frequency in the language the code is from. Since Captain Kidd's code is from the English language, select a paragraph from a book written in English and determine the number of times each letter of the alphabet occurs in that paragraph. Record your findings in Table 7. Then using the fact that the number 8 occurs more often than any other symbol in Captain Kidd's code, substitute the letter that occurred most often in the paragraph you selected for the number 8 in Captain Kidd's code and break the code.

a. Cryptographers have determined that the letter *e* occurs most frequently in the English language. Do your results support this assertion?

b. Using the results of your experiment, what is the empirical probability of an *e* occurring in the English language?

c. In *The Gold Bug* the succession of letters occurring most frequently after the letter *e* is given as: *a, o, i, d, h, n, r, s, t, u, y, c, f, g, l, m, w, b, k, p, q, x, z.* Do the results of your experiment support this succession?

Experiment 7

If you were to roll a single die 100 times, how many 3s would you expect to get? To check your prediction, roll a die 100 times and keep track of each outcome in Table 8 on page 318.

Table 7   Results of observing each letter of the English alphabet in a paragraph from a book written in English

Outcome	Tally	Frequency
a		
b		
c		
d		
e		
f		
g		
h		
i		
j		
k		
l		
m		
n		
o		
p		
q		
r		
s		
t		
u		
v		
w		
x		
y		
z		

Table 8   Results of rolling a single die 100 times

Outcome	Tally	Frequency
1		
2		
3		
4		
5		
6		
		Total = 100

a. How many 3s did you get?

b. Using the results of your experiment, what is the empirical probability of rolling a 3 with a single die?

c. How does your empirical probability compare with the theoretical probability of rolling a 3 with a single die?

d. To what extent would you make decisions based on theoretical probabilities? Explain using your results for this experiment as a basis for your argument.

Illustrate your results for Table 8 with a line graph in the space provided.

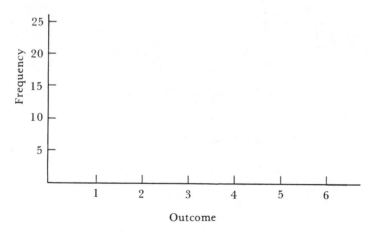

Experiment 8

If you were to roll a pair of dice 100 times, which sum would you expect to get most often? To check your prediction, roll a pair of dice 100 times and keep track of each sum in Table 9.

Table 9   Results of rolling a pair of dice 100 times

Outcome	Tally	Frequency
1		
2		
3		
4		
5		
6		
7		
8		
9		
10		
11		
12		
		Total = 100

a. How many times did you get the sum you predicted?

b. What did you get for the empirical probability of this sum occurring?

c. How does your empirical probability compare with the theoretical probability of this sum occurring?

Illustrate your results for Table 9 with a line graph in the space provided.

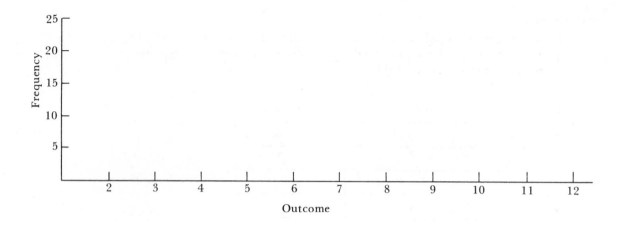

Experiment 9

The true-false answer sheets below refer to different forms of a linear algebra test, and the multiple-choice answer sheets below refer to different forms of a multivariable calculus test. Take the tests by circling either *T* or *F* or *a, b, c,* or *d* for each item of the different forms of the true-false test and the multiple-choice test. Since you do not have the tests, you will have to guess. To see how well you did, check your responses against the answers given at the end of this section. Record the number of correct responses for each form of each test in Table 10.

Form A	*Form B*	*Form C*
1. T　F	1. T　F	1. T　F
2. T　F	2. T　F	2. T　F
3. T　F	3. T　F	3. T　F
4. T　F	4. T　F	4. T　F
5. T　F	5. T　F	5. T　F
6. T　F	6. T　F	6. T　F
7. T　F	7. T　F	7. T　F
8. T　F	8. T　F	8. T　F
9. T　F	9. T　F	9. T　F
10. T　F	10. T　F	10. T　F

Form A	*Form B*
1. a b c d	1. a b c d
2. a b c d	2. a b c d
3. a b c d	3. a b c d
4. a b c d	4. a b c d
5. a b c d	5. a b c d
6. a b c d	6. a b c d
7. a b c d	7. a b c d
8. a b c d	8. a b c d
9. a b c d	9. a b c d
10. a b c d	10. a b c d

*Table 10   Scores for the various forms of the linear algebra
test and the multivariable calculus test*

Number correct on each form of the linear algebra answer sheets			Number correct on each form of the multivariable calculus answer sheets	
Form *A*	Form *B*	Form *C*	Form *A*	Form *B*

a. What was your best score in linear algebra? In multivariable calculus?

b. What should you have scored in linear algebra? In multivariable calculus?

c. Do you think true-false tests or multiple-choice tests give a good indication of what you know?

d. Would you use a right-minus-wrong scoring formula for true-false tests and multiple-choice tests? What formula would you use?

A.   acbccbabca   B.   bcbcaacbcc

A.   TFFTTTTFT   B.   TTFTFTTFFF   C.   FTTFTFFFTT

Summary

The message in this chapter is that probability is primarily a way of thinking about occurrences and that statistics is primarily a way of illustrating those occurrences so that we might think about them clearly. More specifically, probability is a way of trying to second-guess the world, a way of providing some rationale for predicting what is most likely to happen in a given situation. And statistics is a way of organizing and portraying the myriad details of the world in such a way as to facilitate our thinking.

Two aspects of probability were emphasized in this chapter: its theoretical aspect and its empirical aspect. These two aspects of probability do not correlate nearly as well as we would like them to. What ought to have happened and what you found to happen for experiments 7 and 8, for example, were probably two different things. This is not meant to disparage theoretical probabilities. Knowing what ought to happen for a particular event will at least make our predictions for that event correct more often than not.

There is, of course, much more to probability than theoretical and empirical probabilities. For example, there are sophisticated counting techniques utilizing permutations and combinations for determining theoretical probabilities, and there are ways of applying calculus to probability. These topics far surpass the scope of this text. This chapter should be thought of as an introduction to probability as that sort of mathematics one applies rather naturally to occurrences.

The statistics in this chapter was limited to frequency distributions and bar and line graphs for organizing and portraying the data generated in the experiments. Thus the statistics in this chapter emphasized the pictorial aspect of statistics, which is worthwhile since much of the statistics all of us deal with every day in the media is in this form. Concepts such as *mean, median, mode,* and *standard deviation,* which illustrate the numerical side of statistics, could have been introduced as well, but to have done so would have been artificial since none of these concepts would have clarified the mathematical dynamics of the experiments.

References

Fitzgerald, W. M., *et al. Laboratory Manual for Elementary Mathematics.* Prindle, Weber & Schmidt, Boston, 1969.

Oakland County Mathematics Project Staff. *Taking Chances.* The McKay Press, Midland, Michigan, 1971.

Poe, Edgar Allan. "The Gold Bug," *The Complete Tales and Poems of Edgar Allan Poe,* Random House, New York, 1938.

14

Metric measure

More than ever before in this nation's history there is reason to believe that the United States is going to change from its traditional system of measure to the metric system of measure. The National Bureau of Standards puts it this way: The United States is going metric of its own accord at such a rate as to make it just a question of time before the country will be entirely metric. Putting it another way, the day is coming when every American will have to be able to "think metric," that is, be able to estimate in terms of meters, liters, and grams without having to translate from an estimate made first in terms of feet, quarts, and pounds. Metric units are already the basis for nearly all our scientific work, they are used exclusively in our chemical and pharmaceutical industries, and Reddi Kilowatt lights every American home. The major objective of this chapter is to develop the ability to "think metric." A complementary objective is to develop skill in measuring with metric units.

1. The prehistoric system of measure

Thousands of years ago there were few standard units of measure. However, this did not stop ancient people from measuring, and it was only natural to "measure" things in relation to themselves. This was especially easy with lengths. Here are some examples of cave-dweller units of measure for length.

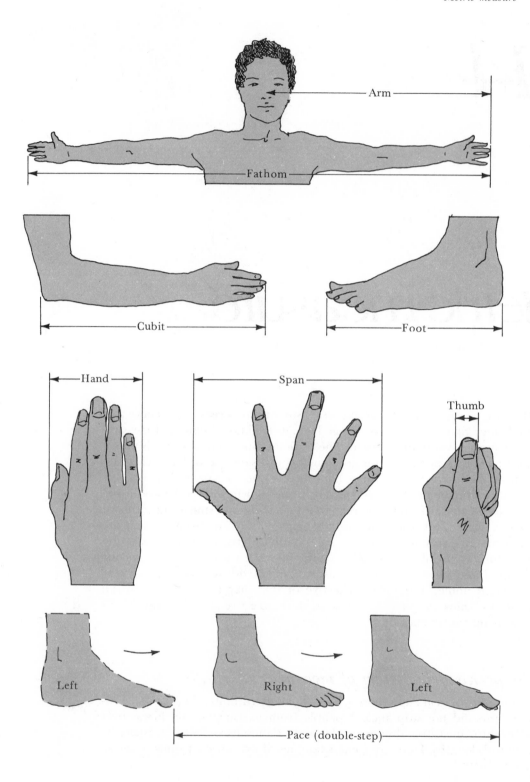

1. Select six distances in your immediate vicinity that range from short to long and describe them in Table 1. For each of the six distances, choose a cave-dweller unit of measure (thumb, hand, span, foot, cubit, arm, fathom, or pace) and use it to measure the distance to the nearest half-unit. Record the cave-dweller unit you used and the measurement you found with it in Table 1.

Table 1   Lengths of six distances in cave dweller units to the nearest half-unit

	Distance	Cave dweller unit of length used	Length to nearest half-unit
Ex.	Width of this text	Thumb	10
1			
2			
3			
4			
5			
6			

2. Compare your results for Table 1 with someone else's results. Are they the same? Why or why not?

3. Two people measured the height of a door to the nearest span. One got a height of 10 spans, the other a height of 12 spans. Which person was correct? Why?

4. Suppose you wanted 50 cubits of rope, and you asked a friend to get it for you. Would you get the length you wanted? Why or why not?

5. List some of the advantages and disadvantages of measuring with parts of the body compared to measuring with the traditional instruments.

6. Explain the need for a standard unit of length—a unit whose length is fixed—in modern society.

7. In recognition of the need for a standard unit of length in modern society, several have been established. For the world in general the standard unit of length is the *meter*. This distance is approximately 1/10,000,000 of the length of the quadrant of the earth's meridian that runs through Paris.

Until 1960 the meter was defined as the distance between two marks on a particular metal bar made of platinum and iridium. Since 1960, however, the meter has been defined as 1,650,763.73 vacuum wavelengths of the orange radiation emitted under specified conditions by the krypton atom of mass 86. As a natural standard, this definition is known to reproduce the meter represented by the former international prototype to within one part in 100,000,000.

In terms of Archimedes' axiom, what does it mean to say that the meter can be reproduced to within one part in 100,000,000?

Archimedes' axiom

If $\overline{AB}$ and $\overline{CD}$ are any segments such that $m(\overline{AB}) < m(\overline{CD})$, and on $\overline{CD}$ we choose points Q_1, Q_2, Q_3, and so on such that $\overline{CQ_1}, \overline{Q_1Q_2}, \overline{Q_2Q_3}$, and so on are all congruent to $\overline{AB}$, then eventually we will find points Q_n and Q_{n+1} for which $D = Q_n$ or D is between Q_n and Q_{n+1}. We then say that the length of $\overline{CD}$ is n or is between n and $n + 1$, respectively.

2. Estimating lengths

For this section you will need a meter stick and the die-cut centimeter rule in the insert in the back of the book.

1. Measure your cave-dweller units to the nearest centimeter or meter and record your findings in Table 2.

2. Compare your table with a neighbor's table. Are they the same? Why or why not?

Table 2 Lengths of cave dweller units
to the nearest centimeter

Cave dweller unit	Length to nearest centimeter or meter
Thumb	
Hand	
Span	
Foot	
Cubit	
Arm	
Fathom	
Pace	

3. Which of the units in Table 2 would be easy for you to remember as a guide for estimating the metric length of short distances? Of long distances?

4. Will the answer to question 3 be the same for everyone? Why or why not?

5. To be useful, what properties must an estimation guide have?

6. Using your thumb as an estimation guide, estimate the lengths of the following segments to the nearest centimeter. Check your estimates with a centimeter rule. How close were you in each case?

7. Using your span as an estimation guide, draw a segment 50 cm in length. Check the accuracy of your drawing with a centimeter rule. By how much were you off?

8. For each of the following pairs of segments, guess which segment is the longer. Check your guesses with a centimeter rule. Did you guess correctly?

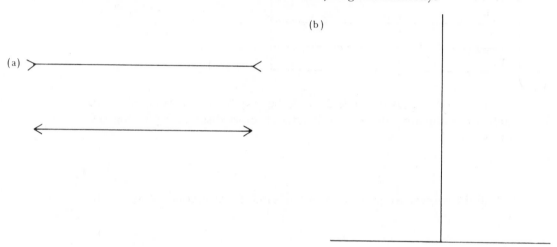

(b)

(a)

9. Using your thumb as an estimation guide, estimate the perimeter of the pentagon below to the nearest centimeter. Check your estimate with a centimeter rule. How close were you?

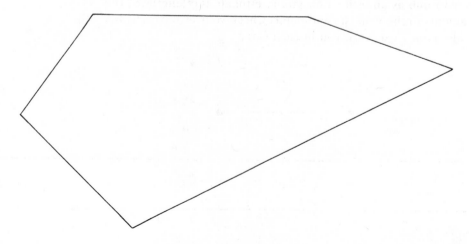

10. Using your thumb, hand, or span as an estimation guide, estimate the flight distance between the cities in Table 3 to the nearest 50 kilometers (km). To do this, use the map in Figure 1 with a scale of 1 centimeter (cm) = 242 km.

Table 3   Estimated flight distances between 13 major cities to the nearest 50 kilometers

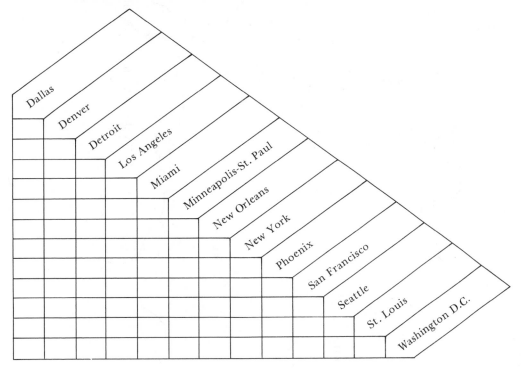

11. Using some part of your body as an estimation guide, estimate your "vital statistics" to the nearest centimeter. Check your estimates with a piece of string. How close were you in each case?

Chest _____ cm

Waist _____ cm

Hips _____ cm

12. a. Copy Table 1 in the first three columns of Table 4 on page 331.

b. Estimate the length of each distance from Table 1 to the nearest centimeter or meter by converting its length in cave-dweller units to its corresponding length in metric units. To do this, refer to Table 2 and *compute.*

c. Check your estimates by measuring each distance with a meter stick.

d. Determine the accuracy of your estimates by computing the difference between your estimate and the measured length of each distance.

e. Record your findings in the second three columns of Table 4.

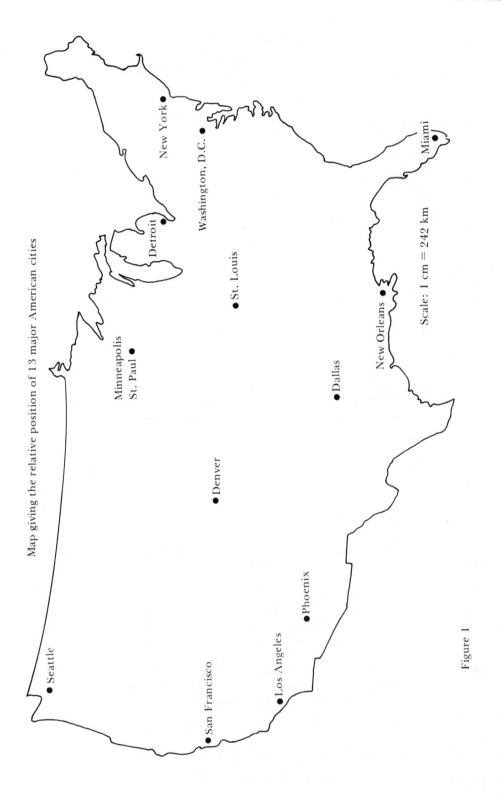

Map giving the relative position of 13 major American cities

Scale: 1 cm = 242 km

Figure 1

Table 4   Estimates of the lengths of six distances in metric units

	Distance	Unit of length used	Length to nearest half-unit	Estimate in metric units	Measure-ment in metric units	Differ-ence
Ex.	Width of this text	Thumb	10	10×2 cm = 20 cm	21.7 cm	1.7 cm
1						
2						
3						
4						
5						
6						

13. Compare your results for Table 4 with someone else's results. To what extent are they the same? Why? To what extent are they different? Why?

14. a. Using some part of your body as an estimation guide, estimate the lengths of the distances in Table 5 to the nearest centimeter, meter, or kilometer.

Table 5   Estimates of the lengths of six more distances in metric units

Distance	Estimate in metric units	Measurement in metric units	Difference
Length of a dollar bill			
Length of a hallway in this building			
Distance from here to your next class			
Height of this building			
Width of a nickel			
Length of the dark blue die-cut strip			

b. Check your estimates with a measurement of each distance. For some distances it will be helpful to know the dimensions of a brick, a paving stone, a floor tile, and the like.

c. Determine the accuracy of your estimates by computing the difference between your estimate and the measured length of each distance.

d. Record your findings in Table 5.

3. Estimating areas

1. Using your thumb or hand an an estimation guide, what are the dimensions of the square in Figure 2 to the nearest centimeter? Check your estimates with a centimeter rule. What is the area of the square to the nearest square centimeter?

Make a square to match the drawing below; then cut it along the dotted lines into four pieces. Assemble these pieces to form a rectangle. What are the dimensions of the rectangle? What is its area? Explain.

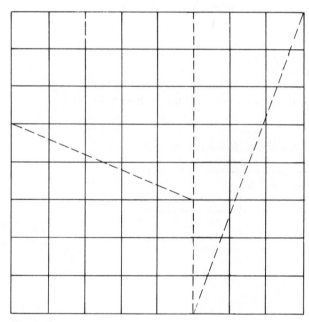

Figure 2

2. Using your thumb as an estimation guide, which of the following regions best represents an area of one square centimeter? Four square centimeters? Use the centimeter rule to check your answers.

3. Using what a square centimeter looks like to you, estimate the area of the following shapes to the nearest square centimeter. Check your estimates with a centimeter rule.

4. Using what a square centimeter looks like to you, estimate the area of the die-cut colored strips from the insert at the back of the book. Check your estimates with a centimeter rule.

5. Estimate the area of the bottom of your foot to the nearest square centimeter. To check your estimate, make an outline of your foot on the graph paper provided in the appendix and count the squares. How close were you?

6. Estimate the surface area of your body. To do this, multiply the area of the bottom of your foot by 100. Knowing the surface area of your hand, how might you determine the accuracy of this method of estimation?

7. Select six regions in the classroom or vicinity such as a blackboard or a door to find their areas.

   a. Describe these regions in Table 6.

Table 6 Estimates of the area of six regions in metric units

	Region	Estimate in metric units	Measurement in metric units	Difference
Ex.	Blackboard	4 m^2	3.6 m $\times$ 1.2 m = 4.3 m^2	0.3 m^2
1				
2				
3				
4				
5				
6				

   b. Using what a square centimeter or meter looks like to you, estimate the area of each region to the nearest of either.

   c. Check your estimates with a measurement of each region.

   d. Determine the accuracy of your estimates by computing the difference between your estimate and the measurement of each region.

   e. Record your findings in Table 6.

8. a. Using what a square centimeter or meter looks like to you, estimate the areas of the regions in Table 7 to the nearest of either.

   b. Check your estimates with a measurement of each region.

   c. Determine the accuracy of your estimates by computing the difference between your estimate and the measurement of each region.

   d. Record your findings in Table 7.

4. Estimating volumes

The primary unit of volume in the metric system is the *liter* which is defined as the volume of a cube one decimeter, or $\frac{1}{10}$ of a meter, on an edge. A reasonably exact die-cut model of a liter is provided in the insert in the back of the book. Another common unit of volume in the metric system is the *milliliter* which is defined as the volume of a cube one centimeter, or $\frac{1}{100}$ of a

Table 7   Estimates of the areas of six more regions in metric units

Region	Estimate in metric units	Measurement in metric units	Difference
Dollar bill			
Quarter			
Hand			
Side of this building			
Dark blue die-cut flat			
Clock			

meter, on an edge. By doing the following activities, you should become fairly adept at estimating the volume of quantities in liters and milliliters.

1. Assemble the die-cut liter box in the insert at the back of the book.

a. Using your thumb as an estimation guide, what are its dimensions to the nearest meter? To the nearest centimeter? Check your estimates with a centimeter rule.

b. What is the volume of the die-cut liter box to the nearest cubic meter?

c. To the nearest cubic centimeter?

2. Using your thumb as an estimation guide, determine which of the following solids best represents a volume of one milliliter. Eight milliliters? Use a centimeter rule to check your answers.

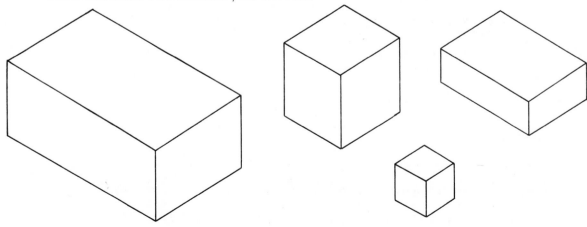

3. Estimate the volume of your body to the nearest liter. To do this, convert your weight from pounds to kilograms by dividing your weight in pounds by 2.2, and use the fact that your body is nearly as dense as water and that one kilogram of water occupies one liter in volume.

4. Select six quantities in the classroom or vicinity such as a trash can or a bookshelf to find their volume.

   a. Describe these quantities in Table 8.

Table 8   Estimates of the volumes of six quantities in metric units

	Quantity	Estimate in metric units	Measurement in metric units	Difference
Ex.	Trash can	20 liters	$3.14 \times 15 \text{ cm}^2 \times 30 \text{ cm} =$ $21195 \text{ cm}^3 = 21.21$	1.2 liters

   b. Using what a liter or milliliter looks like to you, estimate the volume of each quantity to the nearest of either.

   c. Check your estimates with a measurement of each quantity.

   d. Determine the accuracy of your estimates by computing the difference between your estimate and the measurement of each quantity.

   e. Record your findings in Table 8.

5. a. Using what a liter or milliliter looks like to you, estimate the volumes of the quantities in Table 9 to the nearest of either.

   b. Check your estimates with a measurement of each quantity. For those quantities that are irregular in shape, measure their displacement in water.

   c. Determine the accuracy of your estimates by computing the difference between your estimate and the measurement of each quantity.

   d. Record your findings in Table 9.

*Table 9   Estimates of the volumes of six more quantities
          in metric units*

Quantity	Estimate in metric units	Measurement in metric units	Difference
Piece of chalk			
Cup			
Egg			
Bathtub full of water to a height of 10 cm			
Marble			
Telephone book			

5. Estimating mass

The primary unit of weight (or, more specifically, mass) in the metric system is the *gram* which is defined as the weight of one milliliter of water. Another common unit of weight in the metric system is the kilogram, which is 1000 grams, the weight of one liter of water. The die-cut model of a liter that came with this text would therefore weigh one kilogram if it were full of water. By doing the following activities, you should be able to develop some feeling for estimating the weight of objects in grams and kilograms.

1. Select six objects in the classroom or vicinity such as an eraser or a nickel to find their weights.

   a. Describe these objects in Table 10.

Table 10   Estimates of the weights of six objects in metric units

	Object	Estimate in metric units	Measurement in metric units	Difference
Ex.	Eraser	50 g	40 g	10 g
1				
2				
3				
4				
5				
6				

b. Using what a gram or kilogram feels like to you, estimate the weight of each object to the nearest of either.

c. Check your estimates by weighing each object.

d. Determine the accuracy of your estimates by computing the difference between your estimate and the weight of each object.

e. Record your findings in Table 10.

2. a. Using what a gram or kilogram feels like to you, estimate the weight of the objects in Table 11 to the nearest of either.

b. Check your estimates by weighing each object.

c. Determine the accuracy of your estimates by computing the difference between your estimate and the weight of each object.

d. Record your findings in Table 11.

Table 11   Estimates of the weights of six more objects in metric units

Object	Estimate in metric units	Measurement in metric units	Difference
Telephone			
Penny			
Egg			
Bathtub full of water to a height of 10 cm			
Marble			
Telephone book			

6. *The metric system of measure*

The basic units of length, volume, and weight in the metric system of measure are the meter, liter, and gram, respectively. These units are related as follows: The liter is the volume of $\frac{1}{10}$ of a meter cubed, and the gram is $\frac{1}{1000}$ of the weight of a liter of water. The metric units for length, volume, and weight are therefore related in ratios of 10. When we add to "meter," "liter," and "gram" the prefixes *kilo-* (1000), *hecto-* (100), *deka-* (10), *deci-*($\frac{1}{10}$), *centi-* ($\frac{1}{100}$), and *milli-* ($\frac{1}{1000}$), computation within the system reduces to adding a zero or moving a decimal point. The following exercises illustrate the ease of computing in the metric system.

1. Fill in the blanks in the metric conversion tables, Tables 12–16.

Table 12   Conversions from meters to meters

	Kilo-meter	Hecto-meter	Deka-meter	Meter	Deci-meter	Centi-meter	Milli-meter
Kilometer	1	10	100	1000	10,000	100,000	1,000,000
Hectometer	0.1	1					
Dekameter	0.01		1				
Meter	0.001			1			
Decimeter	0.0001				1		
Centimeter	0.00001					1	
Millimeter	0.000001						1

Table 13   Conversions from liters to liters

	Kilo-liter	Hecto-liter	Deka-liter	Liter	Deci-liter	Centi-liter	Milli-liter
Kiloliter	1	10	100	1000	10,000	100,000	1,000,000
Hectoliter	0.1	1					
Dekaliter	0.01		1				
Liter	0.001			1			
Deciliter	0.0001				1		
Centiliter	0.00001					1	
Milliliter	0.000001						1

Table 14   Conversions from grams to grams

	Kilo-gram	Hecto-gram	Deka-gram	Gram	Deci-gram	Centi-gram	Milli-gram
Kilogram	1	10	100	1000	10,000	100,000	1,000,000
Hectogram	0.1	1					
Dekagram	0.01		1				
Gram	0.001			1			
Decigram	0.0001				1		
Centigram	0.00001					1	
Milligram	0.000001						1

Table 15 Conversions from meters to liters

	Kilo-Liter	Hecto-Liter	Deka-liter	Liter	Deci-liter	Centi-liter	Milli-liter
Cubic kilometer							
Cubic hectometer							
Cubic dekameter							
Cubic meter	1						
Cubic decimeter				1			
Cubic centimeter							1
Cubic millimeter							

Table 16 Conversions from liters to grams

	Kilo-gram	Hecto-gram	Deka-gram	Gram	Deci-gram	Centi-gram	Milli-gram
Kiloliter of water							
Hectoliter of water							
Dekaliter of water							
Liter of water	1						
Deciliter of water		1					
Centiliter of water			1				
Milliliter of water				1			

2. The units related in Tables 12–16 are from the metric system of measure. Is it clear that this system is a product of the *mind*? Why or why not?

3. What is meant by the statement that computation in the metric system amounts to little more than adding a zero or moving a decimal point?

4. What is meant by the statement that the metric system is a reflection of decimal arithmetic? In light of question 2, is this surprising? Why or why not?

5. What is meant by the statement that the metric system is decimal all the way?

6. What are the numeric equivalents of the prefixes *kilo-, hecto-, deka-, deci-, centi-,* and *milli-*?

7. Supply the missing prefixes. A liter is the volume of a cubic _____ meter. A gram is the weight of a _____ liter of water, that is, the weight of a cubic _____ meter of water. A liter of water weighs a _____ gram.

8. It has been asserted that our going metric will save as much as three years in the education of a child: "Luckiest, in the face of metrication, would be school children. Since the metric system eliminates most common fractions and complicated measures, they could learn arithmetic in three-fourths of the present time." To what extent do you agree with this statement?

7. *The United States traditional system of measure*

1. Fill in the blanks in Table 17, which is a conversion table for the United States traditional system of measure.

Table 17   Conversions from length to length

	Mile	Furlong	Rod	Fathom	Yard	Foot	Span	Hand	Inch
Mile	1	8	320	880	1760	5280	7040	15840	63360
Furlong	0.125	1							
Rod	0.003125		1						
Fathom	0.001136			1					
Yard	0.0005682				1				
Foot	0.0001894					1			
Span	0.0001420						1		
Hand	0.00006312							1	
Inch	0.00001578								1

2. The units in Table 17 are taken from the United States system of measure. Explain why this system is said to have arisen as a result of women and men of ancient times having applied their *bodies* to quantitative situations.

3. What is meant by the statement that the units from the United States traditional system of measure collide with computation? In light of question 2, is this surprising? Why or why not?

4. American industry uses a "decimal inch" divided in 100ths and 1000ths, and surveyors in this country have a "decimal foot." Nonetheless, the two are incompatible. Why?

5. React to the following statement made by an engineer about our system of measure: "Oh, the brutal waste of the life of holy childhood in learning these endless tables designed by madmen dead and damned!"

6. When England was still using the foot, quart, and pound, an English lord is reported to have said, "We have all the best of it—we can understand the metric units, but foreigners absolutely cannot understand ours. We've got 'em!" Justify his assertion.

7. The following appeared in *World Metric Standardization: An Urgent Issue* (1922). Respond to it in terms of what you have learned from the previous exercises.

Shall It Be This?

Meter . . . . . . Liter . . . . . . Gram

(The 3 units of the decimal metric system, always uniform; known and used thruout the world.)

Or This?

The obsolete jumble of weights and measures with which the English-speaking peoples are still encumbered: yards, fathoms, rods, leagues, perches, links, feet, inches, chains, furlongs, miles, knots, hands, spans, 2 different quarters, quarterns, ounces, minims, drams, grains, scruples, pennyweights, 3 different hundredweights, 2 different tons and 1 tun, 10 different stones, 4 different pounds, 2 different gallons, 2 different quarts, 2 different pints, gills, barrels, many different bushels, many pecks, and so on. Dry measure, wet measure, wine measure, beer measure, avoirdupois, apothecary, troy weight, and so on to an infinity of absurdity.

Summary

Beginning in France in 1791, the metric system of measure swept the world until by 1965 the United States was the only major nation not using or committed to the eventual use of meter, liter, and gram. From an economic standpoint, this meant that the majority of United States products were unsuitable for the world marketplace, a reality that cost this country an estimated $20 billion annually in export losses alone. Concerned with this loss, the 1968 Congress directed the Department of Commerce to explore the feasibility of converting to the metric system of measure. The Department of Commerce in turn directed the National Bureau of Standards to study the problem, and after three years of study the Bureau concluded that the United States was going metric of its own accord at such a rate as to make it just a question of time before the country would become entirely metric. The Bureau therefore recommended that the United States direct its attention to *when* and *how* the country would convert—planned or otherwise.

Planned or not, however, the consequences of a changeover are staggering. Not only will converting to the metric system of measure make obsolete millions of textbooks, cookbooks, maps, rulers, thermometers, gauges, scales, and so on, but industry stands to lose billions in the process. What is the automotive industry to do with production machinery calibrated in inches? Scrap it? What is the lumber industry to do with its stockpiles of pipe, lumber, and brick? Discard them? And what about blueprints and survey records? Are these to be rewritten? To do so would require hundreds of years.

Nonetheless, in spite of the expense and confusion of changing from one system of measure to another, nation after nation has switched to the metric system of measure, which is puzzling since at one time nearly all the world was using the foot, quart, and pound system of measure or a system very much like it. Couldn't meter, liter, and gram just as easily have given way to foot, quart, and pound? The answer is *no*, and the circumstances giving rise to the abandonment of the foot, quart, and pound system of measure are as old as human history.

To make weapons, tools, or garments for themselves, prehistoric people referred to their natural endowment, parts of their bodies. For short distances they used the breadth of a finger or palm; for longer ones, an arm-length or span of their outstretched arms. As these natural references became standardized, they became the units for the foot, quart, and pound system of measure.

The breadth of the thumb was an inch in the English vocabulary. Other units were the finger, hand, foot and arm. Nature, in its orderly way, had arranged that these units be related to each other in multiples of 2, 3, 4—the common ratios of the foot, quart, and pound system of measure. The digit, the breadth of the middle finger, was about three-fourths of an inch. The palm was 4 digits, or 3 inches. The span, the distance covered by the spread hand, was 9 inches. The foot was 12 inches. The cubit, from the elbow to the tip of the middle finger, 18 inches. The yard, measured from the center of the body to the fingertips with the arm outstretched, was 36 inches, 2 cubits, 3 feet, or 4 spans, and so on.

But nature plays odd tricks. Primitive people not only measured with their fingers, hands, feet, and arms, but counted on them as well, which led to the decimal system of arithmetic. Then the shortcomings of the foot, quart, and pound system of measure became apparent.

For nonprecision crafts such as carpentry or bricklaying where subdivisions as crude as $\frac{1}{2}$, $\frac{1}{3}$, and $\frac{1}{4}$ are convenient, the foot, quart, and pound system of measure is suitable. But if measurements as precise as $\frac{1}{256}$ are necessary, computation becomes awkward. Nor does converting these fractions to decimals help: $\frac{1}{2}$ is all right at 0.5, but $\frac{1}{4}$ is 0.25, $\frac{1}{8}$ is 0.125, $\frac{1}{16}$ is 0.0625, and $\frac{1}{3}$ is 0.3333. . . . Nor does decimalizing the units themselves avoid the difficulty since it works for only one unit. Industry uses a "decimal inch" divided into 100ths and 1000ths, and surveyors have a "decimal foot," but the two are incompatible since there are 12 inches, not 10, in a foot.

An even greater problem exists in converting from one kind of unit to another. For example, since the capacity of the United States liquid quart is 57.75 cubic inches, and 27.66 cubic inches of water weighs one pound, avoirdupois, a quart of water weighs 57.75 divided by 27.66 pounds, a time-consuming and cumbersome calculation.

Here then is the fundamental weakness of the foot, quart, and pound system of measure: *It fails to interrelate measurement and computation.* In contrast, the metric system is a reflection of decimal arithmetic. With the meter as its fundamental standard, the liter is the volume of $\frac{1}{10}$ of a meter cubed, and the gram is $\frac{1}{1000}$ of the weight of a liter of water. The meter, liter, and gram are therefore decimal units, related to one another in ratios of 10, and by prefixing these units with *kilo-* (1000), *hecto-* (100), *deka-* (10), *deci-* ($\frac{1}{10}$), *centi-* ($\frac{1}{100}$), and *milli-* ($\frac{1}{1000}$), computation within the system reduces to adding a zero or moving a decimal point. In a world as quantitative as ours, such a useful relationship between measurement and computation was enough to lead to the demise of foot, quart, and pound—units whose relationships collide with computation.

References

"Are Inches, Pints, Pounds on the Way Out in the U.S.?" *U.S. News and World Report,* 71 (1972), 73–74.

Armagnac, A. P. "The New Push for the Metric System: Will You Give Up Pounds, Feet, and Inches?" *Popular Science,* 204 (1969), 54–57.

Asimov, I. *Realm of Measure.* Fawcett World Library, New York, 1960.

Halsey, F. A., and S. S. Dale. *The Metric Fallacy and the Metric Failure in the Textile Industry.* 2 vols. D. Van Nostrand Company, New York, 1904.

Halsey, F. A. *The Metric Fallacy,* 2d ed. The American Institute of Weights and Measures, 1920.

Manchester, H. "Must America Go Metric?" *Readers' Digest* (October 1968), 63–67.

"Metric Bill Finally Voted." *Science News,* 102 (1972), 132.

Oakland County Mathematics Project. *Exploring Linear Measure,* Oakland Schools, 2100 Pontiac Lake Road, Pontiac, Michigan.

Perry, J. *The Story of Standards.* Funk & Wagnalls, New York, 1955.

Shoecraft, P. "The Foot, Quart, and Pound Collide with Computation, U. S. Goes Metric." *School Science and Mathematics,* 651 (1974), 67–68.

U.S. Metric Study. *A Metric America: A Decision Whose Time Has Come.* National Bureau of Standards Special Publication 345 (1971).

15

Geometry

Geometry is the study of form—collections of points in relation to one another. It arose from observations of the environment. The shortest path from one place to another was a straight line. Squares, rectangles, and triangles were apparent in the construction of walls and dwellings. Circles were numerous: the periphery of the sun or moon, the growth rings of a tree, the swelling ripples caused by a pebble cast into a pond. As people began to record these abstractions, they began to discover relationships among them—and geometry became a science.

The geometry of simple plane figures is a fascinating, fruitful, and challenging field of mathematics. The triangle and its associated points, lines, and circles has led to more than 10,000 discoveries! Moreover, new discoveries occur almost daily. Research in this field requires very little in the way of prior mathematical knowledge, and the field is sufficiently rich to offer anyone with a little perseverance a fair chance of making an original contribution. The objective of this chapter is to illustrate how to go about making geometric discoveries.

1. *Inductive reasoning*

The most powerful tool for making geometric discoveries is *induction*—reasoning from the particular to the general. The procedure is to examine specific cases of a geometric "given" in hopes of detecting a geometric relationship other than the given that is common to the cases. In most instances this relationship will be a geometric discovery, something that could be shown to exist in *every* instance of the givens. In the following exercises, some particularly interesting and important geometric relationships can be discovered through induction.

1. *Pascal's "mystic hexagram"* For the inscribed hexagons labeled *ABCDEF* in Figure 1, extend sides $\overline{AB}$ and $\overline{DE}$ until they intersect. Similarly, extend sides $\overline{BC}$ and $\overline{EF}$ and sides $\overline{CD}$ and $\overline{AF}$ until they intersect. Make a conjecture about the points of intersection of the three pairs of opposite sides of a hexagon inscribed in a circle.

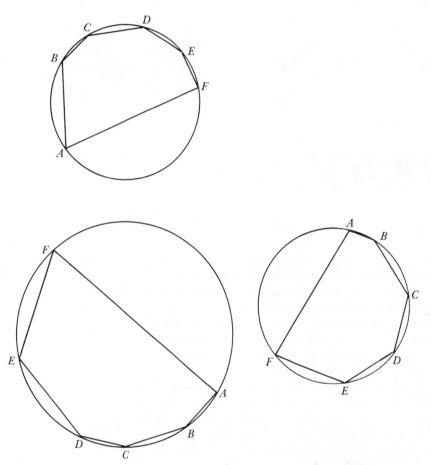

Figure 1

2. Measure sides $\overline{AX}$, $\overline{BX}$, and $\overline{CX}$ of the triangles inscribed in Figure 2 to the nearest millimeter and record your measurements in Table 1. The expressions $m(\overline{AX})$, $m(\overline{BX})$, and $m(\overline{CX})$ in the table represent the measures of the segments $\overline{AX}$, $\overline{BX}$, and $\overline{CX}$, respectively, which make up the sides of the triangles. Complete the table by performing the indicated operation.

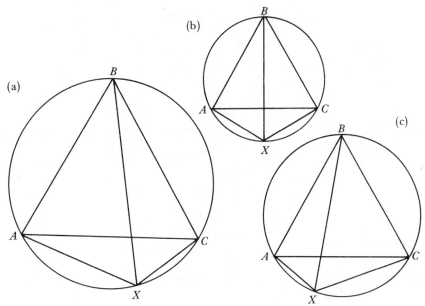

Figure 2

Table 1   Lengths of $\overline{AX}$, $\overline{BX}$, and $\overline{CX}$ to nearest millimeter

	$m(\overline{AX})$	$m(\overline{CX})$	$m(\overline{BX})$	$m(\overline{AX}) + m(\overline{CX})$
a				
b				
c				

The three triangles labeled *ABC* in Figure 2 are *equilateral* triangles. Make a conjecture about a segment that intersects a side of an inscribed equilateral triangle and connects a vertex of the triangle with a point of the circumscribed circle. Draw another such segment in one of the circles of Figure 2 and check your conjecture.

3. *Ptolemy's theorem* For the inscribed quadrilaterals labeled *ABCD* in Figure 3, measure sides $\overline{AB}$, $\overline{BC}$, $\overline{CD}$, and $\overline{AD}$ and diagonals $\overline{AC}$ and $\overline{BD}$ to the nearest millimeter, and record your measurements in Table 2. Complete the table by performing the indicated operations.

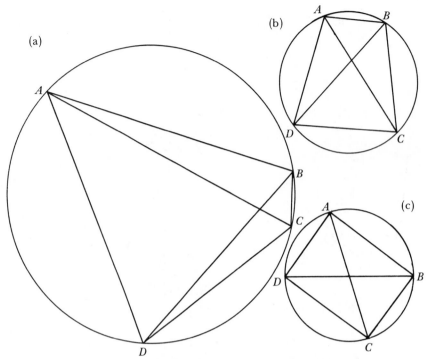

Figure 3

Table 2   Lengths of $\overline{AB}$, $\overline{BC}$, $\overline{CD}$, $\overline{AD}$, $\overline{AC}$, and $\overline{BD}$ to nearest millimeter

	$m(\overline{AB})$	$m(\overline{CD})$	$m(\overline{AD})$	$m(\overline{BC})$	$m(\overline{AC})$	$m(\overline{BD})$	$m(\overline{AB}) \times m(\overline{CD})$ $+ m(\overline{AD}) \times m(\overline{BC})$	$m(\overline{AC}) \times m(\overline{BD})$
a								
b								
c								

Make a conjecture about the sides and diagonals of an inscribed quadrilateral. Draw another inscribed quadrilateral in one of the circles of Figure 3 and check your conjecture.

4. Measure sides $\overline{AB}$, $\overline{BC}$, and $\overline{AC}$, altitude $\overline{BP}$, and median $\overline{AM}$ for each triangle in Figure 4 to the nearest millimeter, and record your measurements in Table 3.

The three triangles labeled *ABC* in Figure 4 are *similar* to one another. Make a conjecture about corresponding parts of similar triangles.

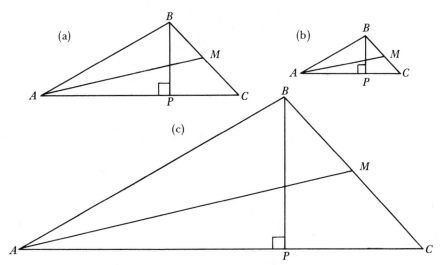

Figure 4

Table 3 *Lengths of $\overline{AB}$, $\overline{BC}$, $\overline{AC}$, and $\overline{BP}$ to nearest millimeter*

	$m(\overline{AB})$	$m(\overline{BC})$	$m(\overline{AC})$	$m(\overline{AM})$	$m(\overline{BP})$
a					
b					
c					

5. Measure parts $\overline{AB}$, $\overline{BC}$, $\overline{AP}$, $\overline{BP}$, and $\overline{CP}$ for each triangle in Figure 5 to the nearest millimeter, and record your measurements in Table 4. Complete the table by performing the indicated operations.

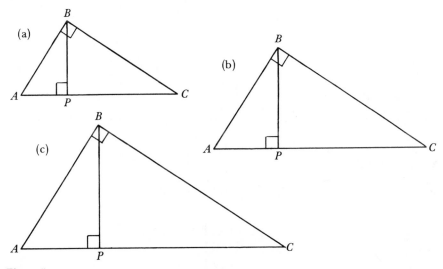

Figure 5

Table 4   Lengths of $\overline{AB}$, $\overline{BC}$, $\overline{AP}$, $\overline{BP}$, *and* $\overline{CP}$ *to nearest millimeter*

	$m(\overline{AB})$	$m(\overline{BC})$	$m(\overline{AP})$	$m(\overline{BP})$	$m(\overline{CP})$	$\dfrac{m(\overline{AB})}{m(\overline{BC})}$	$\dfrac{m(\overline{AP})}{m(\overline{BP})}$	$\dfrac{m(\overline{BP})}{m(\overline{CP})}$
a								
b								
c								

The three triangles labeled *ABC* in Figure 5 are *right* triangles. The three segments labeled $\overline{BP}$ are *altitudes* for these triangles. Make a conjecture about a right triangle *ABC* and the two right triangles *ABP* and *BCP* formed by the altitude $\overline{BP}$ from the vertex of the right angle of triangle *ABC*.

6. Measure segments $\overline{AP}$, $\overline{PB}$, $\overline{CP}$, and $\overline{PD}$ for each circle in Figure 6 to the nearest millimeter, and record your measurements in Table 5. Complete the table by performing the indicated operations.

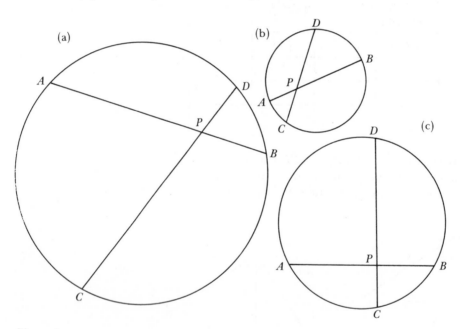

Figure 6

Table 5   Lengths of $\overline{AP}$, $\overline{PB}$, $\overline{CP}$ *and* $\overline{PD}$ *to nearest millimeter*

	$m(\overline{AP})$	$m(\overline{PB})$	$m(\overline{CP})$	$m(\overline{PD})$	$m(\overline{AP}) \times m(\overline{PB})$	$m(\overline{CP}) \times m(\overline{PD})$
a						
b						
c						

A *chord* is a segment whose end points lie on a circle. Segments $\overline{AB}$ and $\overline{CD}$ in Figure 6 are chords. Make a conjecture about intersecting chords. Draw two additional intersecting chords in one of the circles in Figure 6 and check your conjecture.

7. Measure segments $\overline{AP}$, $\overline{BP}$, and $\overline{CP}$ for each semicircle of Figure 7 to the nearest millimeter, and record your measurements in Table 6. Complete the table by performing the indicated operations.

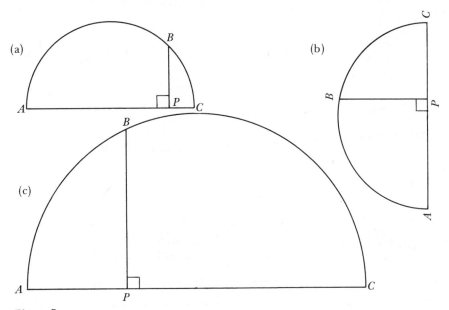

Figure 7

Table 6   *Lengths of* $\overline{AP}$, $\overline{BP}$, *and* $\overline{CP}$ *to nearest millimeter*

	$m(\overline{AP})$	$m(\overline{CP})$	$m(\overline{BP})$	$m(\overline{AP}) \times m(\overline{CP})$	$m(\overline{BP})^2$
a					
b					
c					

Make a conjecture about a perpendicular erected in a semicircle. Erect another perpendicular in one of the semicircles of Figure 7 and check your conjecture.

8. Measure segments $\overline{AB}$, $\overline{AC}$, $\overline{AD}$, and $\overline{AE}$ for each circle of Figure 8 to the nearest millimeter, and record your measurements in Table 7. Complete the table by performing the indicated operations.

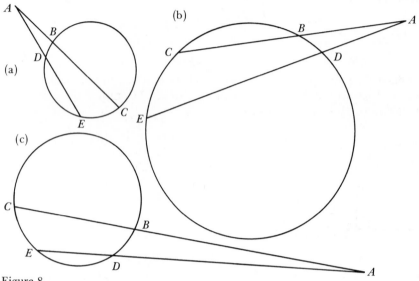

Figure 8

Table 7   Lengths of $\overline{AB}$, $\overline{AD}$, $\overline{AC}$, and $\overline{AE}$ to nearest millimeter

	$m(\overline{AB})$	$m(\overline{AC})$	$m(\overline{AD})$	$m(\overline{AE})$	$m(\overline{AB}) \times m(\overline{AC})$	$m(\overline{AD}) \times m(\overline{AE})$
a						
b						
c						

A *secant* is a line that intersects a circle in two points. Lines $\overleftrightarrow{AC}$ and $\overleftrightarrow{AE}$ in Figure 8 are secants. Make a conjecture about secants that intersect. Draw another pair of intersecting secants and check your conjecture.

9. Measure segments $\overline{AB}$, $\overline{AC}$, and $\overline{AT}$ for each circle of Figure 9 to the nearest millimeter, and record your measurements in Table 8. Complete the table by performing the indicated operations.

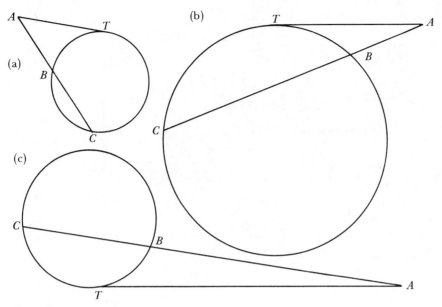

Figure 9

Table 8   *Lengths of $\overline{AB}$, $\overline{AC}$, and $\overline{AT}$ to nearest millimeter*

	$m(\overline{AB})$	$m(\overline{AC})$	$m(\overline{AT})$	$m(\overline{AB}) \times m(\overline{AC})$	$m(\overline{AT})^2$
a					
b					
c					

A *tangent* is a line in the plane of a circle that intersects the circle in exactly one point. The lines through points A and T in Figure 9 are tangents. Make a conjecture about secants and tangents that intersect. Draw another secant and tangent that intersect and check your conjecture.

10. *Menelaus' theorem* Measure segments $\overline{AD}$, $\overline{DB}$, $\overline{BE}$, $\overline{EC}$, $\overline{CF}$, and $\overline{FA}$ for each triangle of Figure 10 to the nearest millimeter, and record your measurements in Table 9. Complete the table by performing the indicated operations.

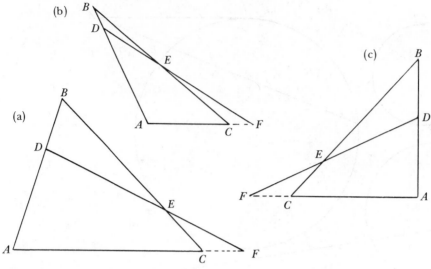

Figure 10

Table 9   Lengths of $\overline{AD}$, $\overline{DB}$, $\overline{BE}$, $\overline{EC}$, $\overline{CF}$, *and* $\overline{FA}$ *to nearest millimeter*

	$m(\overline{AD})$	$m(\overline{DB})$	$m(\overline{BE})$	$m(\overline{EC})$	$m(\overline{CF})$	$m(\overline{FA})$	$\dfrac{m(\overline{AD})}{m(\overline{DB})} \times \dfrac{m(\overline{BE})}{m(\overline{EC})} \times \dfrac{m(\overline{CF})}{m(\overline{FA})}$
a							
b							
c							

A line that intersects in different points two or more lines lying in the same plane is called a *transversal*. The lines through points D, E, and F in Figure 10 are transversals. Make a conjecture about the transversal that intersects the sides of a triangle. Draw a transversal for another triangle and check your conjecture.

11. *Ceva's theorem* Measure segments $\overline{AD}$, $\overline{DB}$, $\overline{BE}$, $\overline{EC}$, $\overline{CF}$, and $\overline{FA}$ for each triangle of Figure 11 to the nearest millimeter, and record your measurements in Table 10. Complete the table by performing the indicated operations.

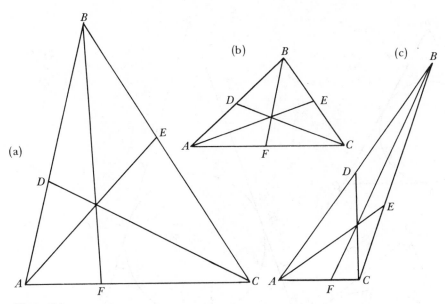

Figure 11

Table 10   *Lengths of* $\overline{AD}$, $\overline{DB}$, $\overline{BE}$, $\overline{EC}$, $\overline{CF}$, *and* $\overline{FA}$ *to nearest millimeter*

	$m(\overline{AD})$	$m(\overline{DB})$	$m(\overline{BE})$	$m(\overline{EC})$	$m(\overline{CF})$	$m(\overline{FA})$	$\dfrac{m(\overline{AD})}{m(\overline{DB})} \times \dfrac{m(\overline{BE})}{m(\overline{EC})} \times \dfrac{m(\overline{CF})}{m(\overline{FA})}$
a							
b							
c							

   Lines that lie in the same plane and intersect in a single point are said to be *concurrent*. Lines $\overleftrightarrow{AE}$, $\overleftrightarrow{BF}$, and $\overleftrightarrow{CD}$ of Figure 11 are concurrent. Make a conjecture about concurrent lines that join the vertices of a triangle to its sides. Draw a triangle and three concurrent lines that join its vertices to its sides and check your conjecture.

12. *Gergonne point* For the circumscribed triangles labeled *ABC* of Figure 12, connect *A* to *L*, *B* to *M*, and *C* to *N*. Make a conjecture about the lines that join the vertices of a triangle to its points of tangency with an inscribed circle.

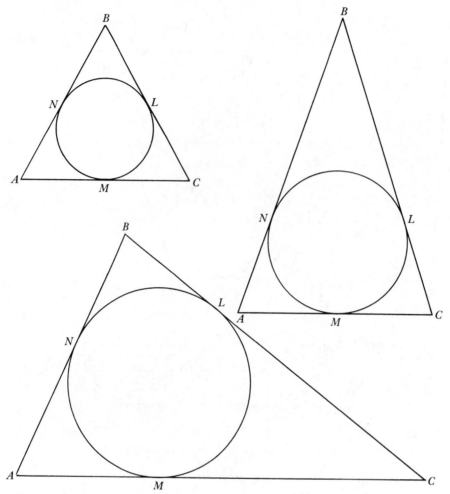

Figure 12

13. Measure angles 1–5 of Figure 13 to the nearest degree and record your measurements in Table 11. The expressions $m(\angle 1)$–$m(\angle 5)$ in the table represent the measures of angles 1–5, respectively.

Table 11 Measure of angles 1 through 5 to the nearest degree

$m(\angle 1)$	$m(\angle 2)$	$m(\angle 3)$	$m(\angle 4)$	$m(\angle 5)$

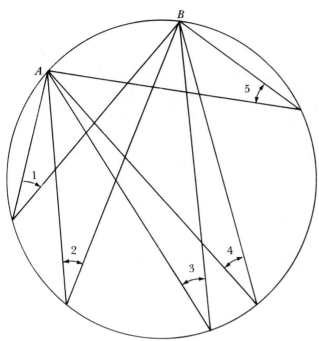

Figure 13

An *inscribed angle* is an angle inside a circle whose vertex lies on the circle. Angles 1–5 are inscribed angles. Each of these angles *intercepts,* or cuts off, arc *AB*. Make a conjecture about inscribed angles that intercept the same arc. Draw another inscribed angle that intercepts arc *AB* to see if its measure agrees with your conjecture.

14. Measure angles 1–5 of Figure 14 to the nearest degree, and record your measurements in Table 12.

Table 12 Measure of angles 1
through 5 to the
nearest degree

$m(\angle 1)$	$m(\angle 2)$	$m(\angle 3)$	$m(\angle 4)$	$m(\angle 5)$

Make a conjecture about inscribed angles that intercept a semicircle. Draw another inscribed angle that intercepts a semicircle to see if its measure agrees with your conjecture.

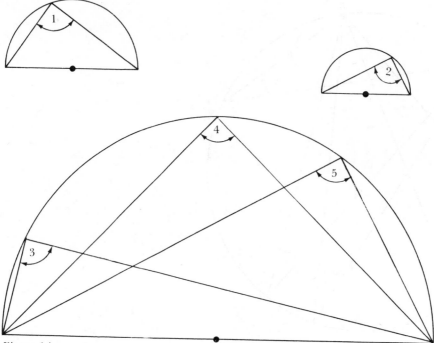

Figure 14

2. *Deductive reasoning*

In the preceding exercises you observed a number of interesting relation-ships between geometric figures. You also made conjectures on the basis of your observations. In this section you will work toward relating the conjec-tures themselves through what is called *deduction*, or deductive reasoning. The work "deductive" is a derivative of the word "deduce," which means to reason necessarily from a set of initial premises. In the following exercises, certain conjectures from your previous work are to be deduced as conse-quences of other conjectures from your previous work. You should begin to see how geometric relationships relate to one another.

EXAMPLE   Deduce the conjecture that $m(\overline{AP}) \times m(\overline{PB}) = m(\overline{PD})^2$ for Figure 15, given that $m(\overline{AP}) \times m(\overline{PB}) = m(\overline{CP}) \times m(\overline{PD})$ in Figure 16.

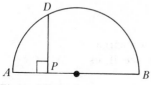

Figure 15

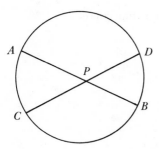

Figure 16

Proof: Complete the circle for Figure 15 and extend $\overline{DP}$ to point C, as shown in Figure 17.

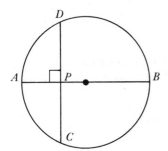

Figure 17

Since $\overline{AB}$ is a diameter in this figure, $m(\overline{PD}) = m(\overline{CP})$, and by substituting $m(\overline{PD})$ for $m(\overline{CP})$ in $m(\overline{AP}) \times m(\overline{PB}) = m(\overline{CP}) \times m(\overline{PD})$, we obtain $m(\overline{AP}) \times m(\overline{PB}) = m(\overline{PD})^2$.

1. Deduce $m(\overline{AP}) \times m(\overline{PB}) = m(\overline{CP}) \times m(\overline{PD})$ for Figure 18, given that inscribed angles which *subtend* the same arc of a circle are equal and that corresponding parts of similar triangles are in proportion.

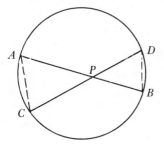

Figure 18

2. Deduce $m(\overline{AP})/m(\overline{PB}) = m(\overline{BP})/m(\overline{CP})$ for Figure 19, given that corresponding parts of similar triangles are in proportion.

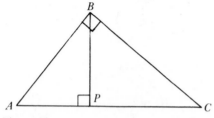

Figure 19

3. Deduce $m(\overline{AP}) \times m(\overline{PB}) = m(\overline{PD})^2$ for Figure 20, given that angles that subtend a semicircle are right angles and that $m(\overline{AP})/m(\overline{PD}) = m(\overline{PD})/m(\overline{PB})$ in Figure 21.

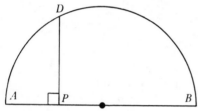

Figure 20

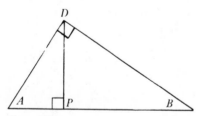

Figure 21

4. Deduce $m(\overline{AC}) \times m(\overline{AB}) = m(\overline{AT})^2$ for Figure 22, given that $m(\overline{AB}) \times m(\overline{AC}) = m(\overline{AD}) \times m(\overline{AE})$ in Figure 23.

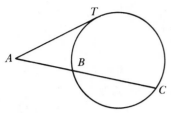

Figure 22

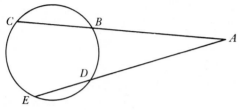

Figure 23

5. Deduce your conjecture about the Gergonne point from Ceva's theorem.

Summary

It was not the intention of this chapter to make the content of geometry apparent. To do so would fill volumes. Rather, we have tried to make apparent a way of viewing geometry, to indicate that geometry is less a collection of abstractions than a book of prose describing the *observable* relationships between collections of points. Seen in this light, geometry is to figures what arithmetic is to numbers: a process and a way of dealing with something, rather than a product or a collection of so-called facts laboriously memorized and soon forgotten. Having completed this chapter, you should feel that geometry, as a *process,* is well within the intellectual capabilities of most individuals, including yourself.

References

Davis, D. R. *Modern College Geometry.* Addison-Wesley Publishing Company, Reading, Massachusetts, 1949.

Eves, H. *A Survey of Geometry.* Allyn and Bacon, Boston, 1972.

Appendix

Model A Twister

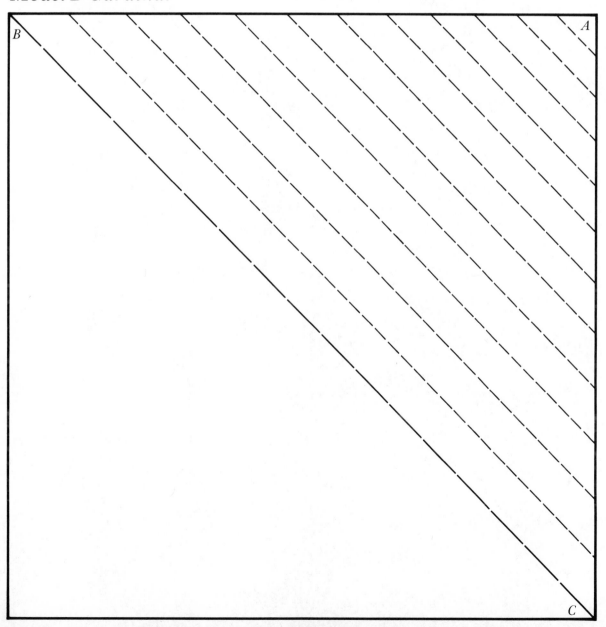

Model B Cardinal

Great twister contest worksheet

Trials–Times

	Student's Name	1	2	3	Average	Place
1						
2						
3						
4						
5						
6						
7						
8						
9						
10						
11						
12						
13						
14						
15						
16						
17						
18						
19						
20						
21						
22						
23						
24						
25						
26						

Construction of a geoboard

A 12-in. piece of $\frac{1}{2}$-in. or $\frac{5}{8}$-in. plywood or particle board will form an excellent base for the 13 × 13 array of nails. Place a piece of graph paper over the board and drive nails at 1-in. intervals. Remove the graph paper, and you are ready to do your assignment.

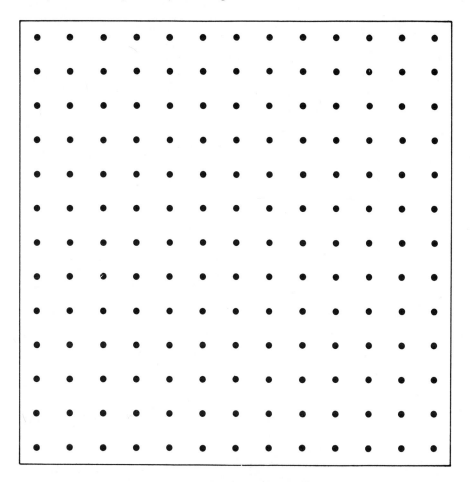

Several suggestions can help you construct a geoboard that will be attractive and easy to work on. First be sure the board is sanded and clean. Second cover it with solid-colored, plastic-coated, adhesive-backed paper, or paint it with semigloss paint. Then after taping graph paper to the board, drive nails through the graph paper. Use lino nails, which are hard to find but excellent because of their rounded smooth heads.

Geosheet workcards

Directions: Cover each geosheet with clear plastic, adhesive-backed shelf paper or laminate with plastic in a heat-press. Use a grease pencil or washable marker pen to do the exercises in this text.

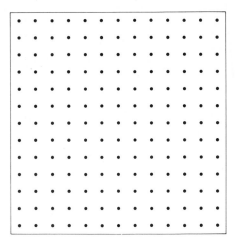

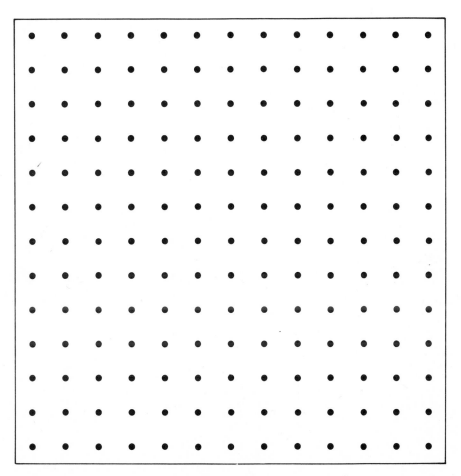

Some interesting geoboard puzzles

1. Go from *A* to *B*, touching all the pegs, in 15 consecutive straight line segments. You may use horizontal, vertical, and diagonal segments, but no line segments may intersect.

2. Go from *C* to *D*, touching all the pegs in 15 consecutive straight line segments. You may use vertical and horizontal but no diagonal segments. No line segments may intersect.

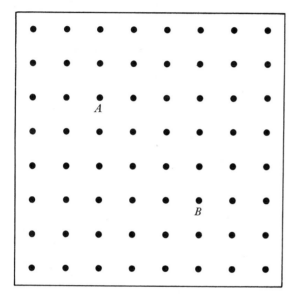

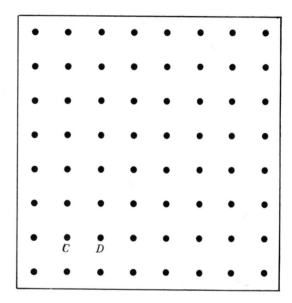

3. Go from *E* to *F* in 17 consecutive straight line segments. The segments may intersect but only at one of the nails.

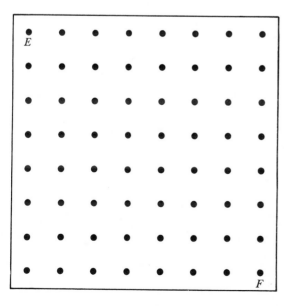

Colored squares game

Number of players   Two to four players.

Materials   Four playing boards, two dice, colored strips, directions.

Directions

1. Players roll dice and take the pair of strips that represents the fraction or any equivalent fraction to the one represented by the dice.

EXAMPLE   If a player rolled a 2 and a 3, the fraction represented is $\frac{2}{3}$, and the player would take a black and a rose *or* a white and an orange *or* an orange and a brown *or* any other equivalent representation of $\frac{2}{3}$ using only two strips.

2. Players place the pair of strips representing the fraction they rolled on their playing board with the strategy of building one-color squares of strips to gain points.

3. Each colored square that is made is worth points equal to its area.

EXAMPLE   A 3 × 3 square (rose) is worth 9 points.

4. Any strip you cannot place on the board must be placed *off* of the playing board and will be counted against the player.

5. The game is over when one player's board is complete or when one player cannot place *either* of the strips on the board.

6. Players subtract the value of the strips they have accumulated off their boards and the number of empty squares on their boards from their points to obtain their scores.

Scoring squares		Penalties	
		Unusable strips	−4
6 × 6 =	36	Empty squares	−2
5 × 5 =	25	Total	−6
4 × 4 =	16		
3 × 3 =	9		
3 × 3 =	9	Total score	
2 × 2 =	4		111
2 × 2 =	4		− 6
2 × 2 =	4		105
2 × 2 =	4		
Total	111		

7. The player with the highest score wins the game.

Colored squares

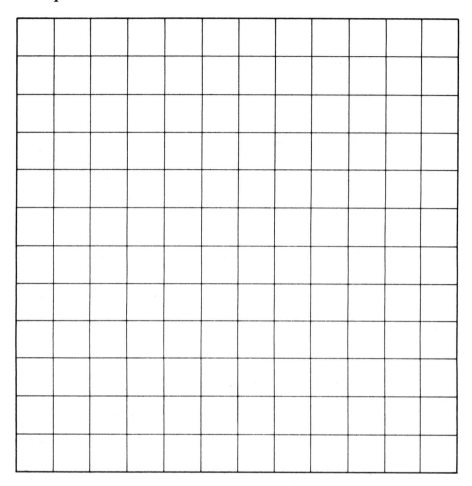

Colored squares

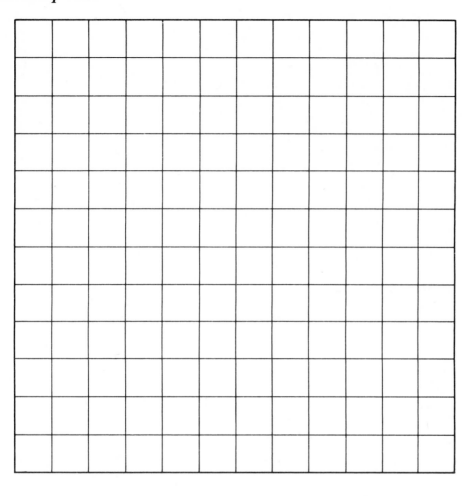

Organizer sheet

Tenths	Hundredths	Thousandths

Number balance assembly instructions

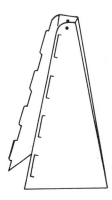

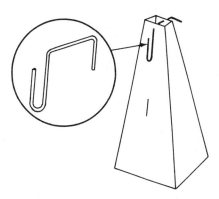

1. Punch out the stand from the insert and fold it along the dotted lines. Fit the tabs into the corresponding slots.

2. Bend a paper clip as shown and place it through the holes in the top of the stand so that the hook is on the front side with the vertical line. Bend down in the back to hold firm.

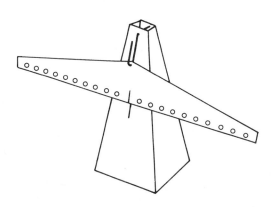

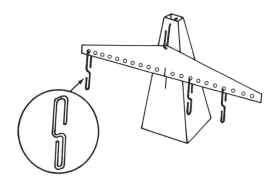

3. Punch out the balance arm, fold it along the dotted line, and hang it on the paper clip hook with the numbered side facing forward. The vertical lines on the balance arm and the stand should be aligned. If not, with a razor knife carefully trim off *small* amounts of the balance arm from the end that is tilted down until the lines are aligned.

4. Bend open several paper clips as shown above. These will be used as weights for the number balance. Check that all the paper clips are the same weight.

Glossary

ACUTE ANGLE   An angle whose measure is between 0 and 90 degrees.

ALGORITHM   A special procedure for solving a certain type of problem.

ASSOCIATIVE   An operation $*$ is associative on a set of elements if for each three elements a, b, c, $(a * b) * c = a * (b * c)$.

AXIS   The coordinate axis used in this textbook. This axis is a vertical or horizontal line along which a coordinate is measured. Usually the axes are perpendicular to each other and the coordinates are equally spaced.

CLOSED   A set and an operation $*$ on the elements of the set are closed if for each pair of elements a and b, $a * b$ equals a member of the set.

COMMUTATIVE   An operation $\#$ is commutative on a set of elements if for each pair of elements a and b, $a \# b = b \# a$.

CONGRUENT   Two shapes are congruent if they are exactly the same size and shape. You can test congruence by seeing if one of the shapes fits exactly on the other.

CONJECTURE   A conjecture is a guess based on incomplete data or a pattern of data. It is usually an inference from intuitive guesswork, based on limited data.

COPLANAR   Sets of points are coplanar if they belong to the same plane.

COUNTEREXAMPLE   An instance or situation that proves that a rule does not hold in all cases.

DEDUCTIVE REASONING   This method makes inferences from accepted principles on the basis of logical relationships.

DISTRIBUTIVE   Operation $*$ is distributive over operation $\#$ if for each three elements a, b, and c, $a * (b \# c) = (a * b) \# (a * c)$.

GEOBOARD   Some type of manipulative device that has arrays of pegs, nails, or holes for representing various mathematical and geometric concepts or relationships.

ESTIMATE   A rough calculation of size, value, etc., based on opinion or judgment.

IDENTITY ELEMENT   A unique element of a set. I is the identity element for an operation $*$ if for every member of the set, $a * I = I * a = a$.

INDUCTIVE REASONING   Drawing conclusions from several known cases. This is a child's basic means of making decisions about how the world works.

ISOCELES TRIANGLE   An isoceles triangle has two sides that are the same length.

ISOMORPHIC   One system is isomorphic to another if the only difference between the systems is the names of the elements and the names of the operations.

OBTUSE ANGLE   An angle whose measure is between 90 and 180 degrees.

ORIGIN   The point of intersection of the vertical and horizontal axes.

PARALLEL   Two lines in the same plane are parallel if, when extended indefinitely, they will not intersect.

PATTERN   A pattern is a sequence or design that has some set rule for successive members.

PERPENDICULAR   Two lines that intersect are perpendicular if they form 90° angles.

PRIME NUMBER   A whole number is prime if it is divisible only by itself and one.

RATIO   The ratio of two numbers is expressed as their quotient. Ratio indicates the relative sizes of two numbers.

RATIONAL NUMBER   Any number that can be expressed as an integer or a quotient of integers.

RECIPROCAL   The number a is the reciprocal of the number b when a times b equals one.

RIGHT TRIANGLE   Any triangle with one angle of 90°.

SET   A collection of things, including abstract ideas, such as numbers.

SIMILAR   Two shapes are similar if their corresponding parts are of the same proportion.

SYSTEM   A set of quantities having some common property. A set of principles concerned with a central objective.

TABLE   A systematic listing of results already worked out that condenses and organizes data.

UNIQUENESS   This is a very important idea in mathematics, and it means that there is *exactly* one result for an operation or relationship.

Selected answers

Chapter 1 Generating mathematical ideas

SECTION 1 Answers will vary.

4. By reversing the fold of the rotor blades, you reverse the direction of the rotation. You can slow the rotation by folding the rotor blades perpendicular to the body of the twister.

SECTION 2 Answers will vary.

1. A common conclusion is: The larger the surface area and the lighter the weight, the longer the cardinal will remain aloft.

SECTION 3 Answers will vary.

2. A common response is a ratio of 2 to 1. Normally a quick estimate with the string produces a loop about 2 times bigger than your actual waist.

3. (a) By measuring with a container the size of a 200-ml cup, the ratio is usually around 20 to 1.
(d) To start this experiment, you must have a hypothesis about what you are researching. It might be: Brand *A* is preferred by more people than Brand *B* in a taste test. Perhaps you can determine whether expensive popcorn is "better" than cheap popcorn.

4. The ratio should be between 2.5 and 3.7 for individuals and 3.0 and 3.3 for the group.

Chapter 2 Patterns and puzzles

SECTION 1

1. 9 vertices, 15 segments, 7 interior regions

3. Networks you design will vary, but vertices plus regions minus one will always equal segments.

SECTION 3

1. One possible conjecture: The sum of consecutive odd integers beginning with one is equal to the square of the number of integers in the series.

3. Add the last two addends of one series to get the last addend of the next series.

SECTION 5

2. (a) 13, 31, 15, 51, 35, 53
$$13 + 31 + 15 + 51 + 35 + 53 = 198$$
$$1 + 3 + 5 = 9$$
$$\frac{198}{9} = 22$$

3. (c) Yes, the procedure does not work for numbers whose digits are all 9's.
    Example   99
$$9 + 9 = 18$$
$$1 + 8 = 9$$

$$9\overline{)9} \quad \frac{1}{}$$
$$\frac{9}{0}$$

SECTION 6

2.

16	3	2	13
5	10	11	8
9	6	7	12
4	15	14	1

SECTION 7

2. Clark is the carpenter, Goldman is the painter, and Kondichook is the plumber.

SECTION 8

1. The average rate will not depend on the distance up the hill; hence, you can "choose" a distance of, say, 3 mi up and 3 mi down the hill. This gives an average rate of $1\frac{1}{2}$ mph.

2. 4 minutes to pass

Chapter 3 Geopatterns

SECTION 1

1. Triangle has 3 sides. Quadrilateral has 4 sides. Pentagon has 5 sides. Hexagon has 6 sides. Octagon has 8 sides.

4. The diagonal of a rectangular region divides it in half. $E = 3$.

6. $A = 6$ $B = 10$ $C = 3\frac{1}{2}$
    $D = 3$ $E = 3\frac{1}{2}$ $F = 9$
    $G = 9$

9. There are many with the minimum area of $\frac{1}{2}$ when we restrict the vertices to nails.

10. The areas of these figures are the same.

SECTION 2

1. (a) 1   (b) 4   (c) 9
   (d) 16   (e) 25   (f) 36
   (g) 49   (h) 64

2. Side length × side length = area, or side length = $\sqrt{\text{area}}$.

4. (a) 2   (b) 8   (c) 18
   (d) 32   (e) 50   (f) 72

Conjecture: The area increases in multiples of four after the first increase of 6. The sequence of #4 is one-half the corresponding entries of the sequence in #3.

6. (a) 2   (b) 5   (c) 10
   (d) 17   (e) 26   (f) 37
   (g) 50   (h) 65

SECTION 3

2. (a) 3   (b) 6   (c) 10
   (d) 15   (e) 21   (f) 28
   (g) 36   (h) 45

The sequence increases by the next whole number, beginning with 3.

5. Sequence A: (a) 1 (b) 2 (c) 3 (d) 4 (e) 5 (f) 6
   Sequence B: (a) 2 (b) 4 (c) 6 (d) 8 (e) 10 (f) 12

7. The area of a triangle is one-half the product of the base times the height: $A = \frac{1}{2} bh$.

8. (a) 10   (b) 10   (c) 10
   (d) 10   (e) 10   (f) 10
   (g) 10

When the base and the height of a triangle remain the same, the area remains the same.

14. *Table 2   Squares on triangles*

	Area of square on one leg	Area of square on second leg	Area of square on hypotenuse
A	1	4	5
B	4	9	13
C	9	16	25
D	16	25	41
E	25	36	61

SECTION 4

2. (a) $\sqrt{4}$ (b) $\sqrt{5}$ (c) $\sqrt{8}$
   (d) $\sqrt{13}$ (e) $\sqrt{20}$ (f) $\sqrt{29}$
   (g) $\sqrt{40}$

The numbers under the square-root sign increase by the consecutive odd numbers.

SECTION 5

1. $D = 7 + \sqrt{13}$ $E = 8 + \sqrt{8} + \sqrt{2} = 8 + 2\sqrt{2} + \sqrt{2} = 8 + 3\sqrt{2}$
   $F = 5 + \sqrt{10} + \sqrt{5}$

3. (a) $3 + \sqrt{5}$ (b) $4 + 2\sqrt{2}$ (c) $5 + \sqrt{13}$
4. (a) $2 + 2\sqrt{5}$ (b) $4 + 4\sqrt{5}$.

SECTION 6
1. (a) Triangle A: $\sqrt{2}, \sqrt{5}, 3$
       Triangle B: $\sqrt{2}, \sqrt{5}, 3$
   (b) They are the same.
   (c) $\frac{3}{2}, \frac{3}{2}$

3. Congruent shapes are the same size and shape. They have the same area, and the length of their sides are the same, respectively.

5. There are many ways.

SECTION 7
1. (a) Triangle A: $1, 2, \sqrt{5}$
       Triangle B: $2, 4, 2\sqrt{5}$
   (b) Sequence B is twice sequence A.

5. Length of sides of *small* 1, 2
   Length of sides of *medium* 2, 4
   Length of sides of *large* $\sqrt{5}, 2\sqrt{5}$

8. Area: *small* = 3   *small* = 3   *large* = 6
   Yes, $3 + 3 = 6$.
   Yes, they are similar.

Chapter 4 Finite systems

SECTION 1
4. Samples: $E * C = A$ $I * B = B$ $C * C = I$
               $I * A = A \cdot$ $B * I = B$ $I * I = I$

SECTION 2
1. $R_1 \# F_A = \underline{F_B}$ $F_B \# F_C = \underline{R_2}$
   $F_A \# F_B = \underline{R_2}$ $F_B \# \underline{F_B} = R_0$

6. $(F_A \# R_1) \# R_2 = F_B \# R_2 = F_A$
   $(R_2 \# F_B) \# F_A = F_A \# F_A = R_0$

7. Yes

8. No

SECTION 3
1. (e) No

Chapter 5 Modular Arithmetic

SECTION 1
1. (a) $o \oplus p = g$
       $p \oplus p = k$
       $n \oplus d = p$
       $y \oplus n = b$

   (b) Yes, every pair of elements laid end-to-end, gives as a result an element of the set of strips.

2. (a) (i) $(k \oplus p) \oplus w = n \oplus w = g$
  (b) Yes, $\oplus$ is associative.

4. (a) $b \oplus p = i$
   $\quad o \oplus o = i$
  (b) Yes
  (c) No

SECTION 2

2. Yes, the elements of the two systems are interchangeable.

SECTION 3

2. $i *$ (any element of the set) $= i$. When i is combined with any element from the set under the $*$ operation, the i "annihilates" the other element.

4. Yes; yes; yes; w; no; no

5. $\otimes$ is an operation where you multiply the two numbers, then discard multiples of 12. The remainder is the result and must not be greater than 12. This system is isomorphic to the $\circledast$ system since the two operations behave in the same ways, and the names of the elements of the two systems can be interchanged.

7. For any 3 elements of the system a, b and c, $a * (b \# c) = (a * b) \# (a * c)$.   Yes

9. (c) $\oplus$ is closed.
       $\oplus$ is associative.
       b is the identity element.
       Each element has a $\oplus$ inverse.
  (d) $\otimes$ is closed.
       $\otimes$ is associative.
       g is the $\otimes$ identity element.
       Except for the annihilator, all elements have an inverse.

SECTION 4

1. (a) w (b) g (c) g
   (d) b (e) r (f) w
   (g) Doesn't exist.   (h) g

3. (a) g
  (b) g
  (g) For any elements, a and b, $^-a \otimes {}^-b = a \otimes b$.

5. (a) $k \otimes k = w$
  (b) $r \otimes r = w$
  (c) $r \otimes g = r$
  (d) $k \otimes g = k$
  (h) Cannot do this! No $\otimes$ inverse for b.
  (i) For any two elements a and b, $(a^{-1} \times b) = (a \times b^{-1})^{-1}$.

SECTION 5

1. r; w; k; g; r; w.
3. $w^1 = w$; $w^2 = g$; $w^3 = w$; $w^4 = g$; etc.
White is not a generator.

5. You simply add the exponents.

7. Zero power behaves like the number 1 with the $\otimes$ operation.

SECTION 6

1. (a) *r*
   (b) *r*
   (c) *g*
   (d) *g*
   (e) *k*
   (f) *k*
   (g) *k*
   (h) *k*
   (i) For any two elements *a* and *b*, $a \ominus b = a \oplus {}^-b$.

SECTION 7

1. (a) *r*
   (b) *r*
   (e) *b*
   (f) *b*
   (i) For any two elements *a* and *b*, $a \otimes b^{-1} = a \oslash b$.

Chapter 6 Base-five arithmetic

SECTION 1

2. (a) 3
   (b) 4

4. (a) 8
   (b) 16

6. (a) Two rose and three black
   (b) One light blue and four black and two green

8. (a) 2 light blue strips and 4 units
   (b) 3 light blue strips and 0 units
   (c) 3 light blue strips and 1 unit
   (d) 3 light blue strips and 2 units

13. (a) 1 quare, 3 quint, 4
    (b) 2 quare, 2
    (c) 4 quint
    (d) 4 quare, 4 quint, 3

16. Million, hundred thousands, ten thousands, thousands, hundreds, tens, ones.
Each unit is ten times the previous unit.
Yes, quillion, quare qube, quint qube, qube, quare, quints, ones.

SECTION 2

3. (Table 4, see following page.)

4. (a) Yes
   (b) Yes
   (c) Yes $a + 0 = a = 0 + a$
   (d) No
   (e) Yes

SECTION 3

1. (a) 3   (b) 10*b*
   (c) 20*b*   (d) 13*b*

Table 4 Addition using standard chains

	1	2	3	4	10b	11b	12b	13b	14b	20b	21b	22b
1	2	3	4	10b	11b	12b	13b	14b	20b	21b	22b	23b
2	3	4	10b	11b	12b	13b	14b	20b	21b	22b	23b	24b
3	4	10b	11b	12b	13b	14b	20b	21b	22b	23b	24b	30b
4	10b	11b	12b	13b	14b	20b	21b	22b	23b	24b	30b	31b
10b	11b	12b	13b	14b	20b	21b	22b	23b	24b	30b	31b	32b
11b	12b	13b	14b	20b	21b	22b	23b	24b	30b	31b	32b	33b
12b	13b	14b	20b	21b	22b	23b	24b	30b	31b	32b	33b	34b
13b	14b	20b	21b	22b	23b	24b	30b	31b	32b	33b	34b	40b
14b	20b	21b	22b	23b	24b	30b	31b	32b	33b	34b	40b	41b
20b	21b	22b	23b	24b	30b	31b	32b	33b	34b	40b	41b	42b
21b	22b	23b	24b	30b	31b	32b	33b	34b	40b	41b	42b	43b
22b	23b	24b	30b	31b	32b	33b	34b	40b	41b	42b	43b	44b

3. (a) 3   (b) 13b   (c) 20b
    (d) 12b   (e) 110b   (f) 221b

5. (a) $24b - 12b = 12b$
    (b) $30b - 14b = 11b$
    (c) $34b - 20b = 14b$

SECTION 4

2. (a) $2 \times 3 \times 10b = 110b$
    (b) $3 \times 4 \times 11b = 242b$
    (c) $2 \times 2 \times 3 \times 3 \times 10b = 1210b$
    (d) $2 \times 3 \times 10b \times 10b = 1100b$

4. (a) Yes   (b) Yes   (c) Yes, 1
    (d) No   (e) Yes

SECTION 5

1. (a) Black   (b) White
    (c) Orange   (d) White

3. (a) The number from the left column was larger than the number from the top row.
    (b) 1; any number divided by itself is equal to 1.
    (c) You cannot divide this in this system because the result is not a whole number.
    (d) 14b is odd. We could divide 14b by 2 if we allow a remainder.

SECTION 6

1. (a) All the numbers with *yes* circled are divisible by 2.
    (b) If the sum of the digits of a number is even, the number can be represented with an all-black chain. *Example: 13b; 1 + 3 = 4.*

2. (a) Rose
   (b) Impossible
   (c) Light blue (Answer may vary.)
   (d) Impossible

3. (a) 22*b*   (b) 24*b*
   (c) 30*b*   (d) 44*b*

5. (b) Add alternate digits; ones + quares + quint qubes, etc. Now add the other digits; quints, qubes, quare qubes, etc. Find the difference of these two sums. This difference will be a multiple of 3 expressed in base five.
   (c) (iv) Very difficult to use a rule—best to just divide it out.

Chapter 7 Base-ten arithmetic

SECTION 1

1. (a) 1 dark blue and 4 unit strips
   (b) 1 dark blue and 4 unit strips
   (c) 1 dark blue and 6 unit strips
   (d) 1 dark blue and 7 unit strips

4. (a) 0 flats, 7 strips, 3 units
   (b) 0 flats, 9 strips, 2 units
   (c) 0 flats, 8 strips, 3 units
   (d) 0 flats, 7 strips, 1 unit

7. 1 hundred, 3 tens, 2 ones

9. (a) 202; two hundred two
   (b) 90; ninety
   (c) 484; four hundred eighty-four

SECTION 2

2. (a) Yes   (b) Yes   (c) Yes
   (d) No   (e) Yes

SECTION 3

1. (a) Rose   (b) Light blue
   (c) Dark blue   (d) Yellow

3. (a) 12
   (b) 20
   (c) 33

4. (a) $20 - 8 = 12$
   (b) $32 - 12 = 20$
   (c) $57 - 24 = 33$

5. (a) $28 - 9 = 18$
   (b) $35 - 14 = 21$
   (c) $59 - 37 = 22$

7. Extend the system to include additive inverses (negative whole numbers) which would be the system of integers.
No.

SECTION 4

1. (Table 3)

Table 3   Multiplication in dark blue standard

X	1	2	3	4	5	6	7	8	9	10	11	12
1	1	2	3	4	5	6	7	8	9	10	11	12
2	2	4	6	8	10	12	14	16	18	20	22	24
3	3	6	9	12	15	18	21	24	27	30	33	36
4	4	8	12	16	20	24	28	32	36	40	44	48
5	5	10	15	20	25	30	35	40	45	50	55	60
6	6	12	18	24	30	36	42	48	54	60	66	72
7	7	14	21	28	35	42	49	56	63	70	77	84
8	8	16	24	32	40	48	56	64	72	80	88	96
9	9	18	27	36	45	54	63	72	81	90	99	108
10	10	20	30	40	50	60	70	80	90	100	110	120
11	11	22	33	44	55	66	77	88	99	110	121	132
12	12	24	36	48	60	72	84	96	108	120	132	144

3. (a) 60   (b) 144
   (c) 210   (d) 81

5. (a) Yes
   (b) Yes; $(2 \times 3) \times 4 = 24$ and $2 \times (3 \times 4) = 24$
   (c) Yes; 1
   (d) No
   (e) Yes; $(3 \times 4) = 12 = (4 \times 3)$

SECTION 5

2. (a) Three blacks and one rose
   (b) No stack possible
   (c) Three rose
   (d) No stack possible

3. (Table 4)

SECTION 6

2. (a) Rose
   (b) Impossible
   (c) Black or light blue

4. (d) (i)   Yellow
       (ii) White
       (iii) Orange
       (iv) Pink

Table 4 Find the other strip

Make a two stack using one of these numbers

	1	2	3	4	5	6	7	8	9	10	11	12
1	1											
2	2	1										
3	3		1									
4	4	2		1								
5	5				1							
6	6	3	2			1						
7	7						1					
8	8	4		2				1				
9	9		3						1			
10	10	5			2					1		
11	11										1	
12	12	6	4	3		2						1
13	13											
14	14	7					2					
15	15		5		3							
16	16	8		4				2				
17	17											
18	18	9	6			3			2			
19	19											
20	20	10		5	4					2		
21	21		7				3					
22	22	11									2	
23	23											
24	24	12	8	6		4		3				2
25	25				5							

Select one of these numbers

Chapter 8 Integers

SECTION 1

1. (a) 6*L*   (b) 4*R*   (c) 5*R*
   (d) 6*R*   (e) 6*L*   (f) 5*L*

5. (a) ⁻5   (b) ⁻2   (c) 0
   (d) ⁻4   (e) ⁺4   (f) ⁺3
   (g) ⁺5   (h) ⁺2

8. (a) ⁺7   (b) ⁻54   (c) 0
   (d) ⁺8   (e) ⁻3   (f) ⁺32
   (g) ⁻15   (h) ⁻7

9. Yes

SECTION 2

1. (a) ⁺3   (b) ⁺1   (c) 0
   (d) ⁻9   (e) ⁻4   (f) ⁺7
   (g) ⁻8   (h) 0

3. (a) ⁺6
   (b) ⁻2
   (c) ⁺5

5. (a) ⁻9   (b) ⁻4
   (c) ⁺36   (d) ⁻56

SECTION 3

2. *Table 3   Negative jumps*

Number of jumps	Size of jumps	Landing point
0	⁻3	0
⁺1	⁻3	⁻3
⁺2	⁻3	⁻6
⁺3	⁻3	⁻9
⁺4	⁻3	⁻12
⁺5	⁻3	⁻15
⁺6	⁻3	⁻18

4. (a) ⁺6   (b) ⁻6
   (c) ⁻6   (d) ⁺6

5. Positive times positive is positive. Positive times negative is negative. Negative times positive is negative. Negative times negative is positive.

9. No, there are no multiplicative inverses for each integer.

SECTION 4

2. ⁺3 cannot be reached by using jumps of ⁺2. You will either be too short or too long.

4. It's impossible to take 0 jumps and land on ⁺12.

5. (e) No; 8 ÷ 4 ≠ 4 ÷ 8.

Chapter 9 Rational numbers

SECTION 1

1. $g = \frac{1}{4}$ $k = \frac{2}{4} = \frac{1}{2}$ $r = \frac{3}{4}$
$w = \frac{4}{4} = 1$ $b = \frac{5}{4}$ $o = \frac{6}{4} = 1\frac{1}{2}$
$p = \frac{7}{4} = 1\frac{3}{4}$ $y = \frac{8}{4} = 2$ $n = \frac{9}{4} = 2\frac{1}{4}$
$d = \frac{10}{4} = 2\frac{1}{2}$ $t = \frac{11}{4} = 2\frac{3}{4}$ $i = \frac{12}{4} = 3$

3. (a) 3   (b) $3\frac{3}{4}$
   (c) $3\frac{1}{2}$ (d) 4

5. (a) Six white
   (b) Seven white and a black
   (c) Twenty white and a green
   (d) Thirteen white and a green
   (e) Impossible
   (f) Impossible

8. It is a very cumbersome procedure.

10. Each strip needs to be "white" wide.

SECTION 2

1. $g = \frac{1}{8}$
   $k = \frac{1}{4}$
   $r = \frac{3}{8}$
   $w = \frac{1}{2}$
   etc.

2. $g = \frac{1}{12}$
   $k = \frac{1}{6}$
   $r = \frac{1}{4}$
   $w = \frac{1}{3}$
   etc.

4. (a) k, b (b) o, k (c) k, g
   (d) r, p (e) n, d

6. Their ratios are equal.

8. $g/w, k/y, r/i$

10. (a) $\frac{2}{4}, \frac{3}{6}$ (b) $\frac{4}{6}, \frac{6}{9}$ (c) $\frac{2}{6}, \frac{3}{9}$
    (d) $\frac{6}{8}, \frac{9}{12}$ (e) $\frac{2}{8}, \frac{3}{12}$

12. $\frac{1}{3} + \frac{1}{11} = ?$ No.

14. (a) $\frac{3}{6} + \frac{4}{6} = \frac{7}{6}$

15. (a) 12

17. (a) $\frac{2}{4} + \frac{1}{4} = \frac{3}{4}$ (b) $\frac{4}{6} + \frac{5}{6} = \frac{9}{6} = \frac{3}{2}$
    (c) $\frac{2}{10} + \frac{5}{10} = \frac{7}{10}$ (d) $\frac{9}{12} + \frac{8}{12} = \frac{17}{12}$

18. (a) Yes   (b) Yes   (c) Yes, o
    (d) Yes, if you include negative fractions.   (e) Yes

20. (a) $\frac{2}{4}$
    (b) $\frac{5}{6} - \frac{2}{6} = \frac{3}{6} = \frac{1}{2}$
    (c) $\frac{7}{8} - \frac{4}{8} = \frac{3}{8}$

21. (a) Yes
    (b) No; $(\frac{3}{4} - \frac{1}{4}) - \frac{1}{4} = \frac{1}{4}$; $\frac{3}{4} - (\frac{1}{4} - \frac{1}{4}) = \frac{3}{4}$
    (c) No; $0 - \frac{1}{2} \neq \frac{1}{2} - 0$
    (d) No

23. (a) $\frac{2}{6} = \frac{1}{3}$
    (b) $\frac{2}{15}$
    (c) $\frac{1}{8}$

25. (a) $\frac{3}{32}$
    (b) $\frac{3}{8}$
    (c) $\frac{5}{24}$

27. (a) $\frac{3}{10}$
    (b) $\frac{2}{7}$

28. (a) $\frac{6}{10} \div 2 = \frac{3}{10}$

29. (a) $\frac{2}{8} = \frac{1}{4}$

31. (a) $\frac{2}{5} \times \frac{3}{2} = \frac{6}{10} = \frac{3}{5}$
    (b) $\frac{4}{3} \times \frac{2}{5} = \frac{8}{15}$

33. (a) $\frac{3}{2}$ (b) $\frac{4}{5}$
    (c) $\frac{7}{4}$ (d) $\frac{5}{2}$

SECTION 3

1. (a) 0.25   (b) 0.6
   (c) 0.38   (d) 0.5

3. (a) $\frac{1}{2}$ (b) $\frac{6}{10}$
   (c) $\frac{4}{10}$ (d) $\frac{3}{10}$

5. RULE: If the digit is less than 5, drop it off. If the digit is 5 or greater, add one to the preceding digit.

8. (a) $\overline{)0.\overline{142857}}$
   (b) $1.\overline{3}$
   (c) $0.\overline{5}$

11. (a) 0.6   (b) 0.75
    (c) 0.25   (d) 0.4

12. (a) 2 flats
    (b) 2 flats and one strip

13. (a) 0.338
    (b) 0.928
    (c) 0.682

14. (a) 0.309
    (b) 0.513
    (c) 0.066

16. (a) 0.0046
    (b) 0.0405
    (c) 0.2108

18. (a) 0.005
    (b) 0.00005

SECTION 4

1. (a) 75%   (b) 35%
   (c) 7%   (d) 44%

2. (a) $\frac{60}{100} = \frac{3}{5}$ (b) $\frac{75}{100} = \frac{3}{4}$
   (c) $\frac{32}{100} = \frac{8}{25}$ (d) $\frac{48}{100} = \frac{12}{25}$

3. (a) $3300

4. (a) 63%

5. (a) 60%

6. (a) 52%

SECTION 5

1. (a) 3.2×10
   (b) $4.287 \times 10 \times 10 \times 10$
   (c) $6.042 \times 10 \times 10 \times 10 \times 10 \times 10 \times 10$
   (d) $4.21 \times \frac{1}{10}$

2. (a) 10^3 (b) 10^2 (c) 10^1
   (d) 10^0 (e) 10^{-1} (f) 10^{-2}
   (g) 10^{-3} (h) 10^{-4} (i) 10^{-5}

4. (a) 2.43×10^2 (b) 3.428×10^3
   (c) 9.098×10^6 (d) 2.4×10^{-2}

SECTION 6

1. 1 light blue strip $= \frac{1}{25}$
   1 green unit $= \frac{1}{125}$

2. (a) $0.112b$ (b) $0.414b$
   (c) $0.311b$ (d) $0.031b$

3. (a) 4 quinths, 1 quarth, 2 qubths

4. (a) $0.344b$ (b) $0.343b$
   (c) $0.320b$ (d) $0.010b$

SECTION 7

1. $A \frac{3}{4}$ $B \frac{3}{6}$ $C \frac{2}{5}$

3. (a) All the points lie on the same straight line.
   (b) Yes, they are the same ratio. They are equivalent.
   (d) It passes through other points.
   (e) They would lie on the same line.

5. They form a straight line passing through the origin.

10. (a) $\frac{3}{4}$ is larger than $\frac{2}{3}$. The slope of the rubber band to $\frac{3}{4}$ is "steeper" than the slope of the rubber band to $\frac{2}{3}$.

11. (a) $\frac{6}{5}$
    (b) $\frac{5}{3}$
    (c) $\frac{5}{6}$

13. $\frac{1}{6}$; $\frac{1}{2} + \frac{1}{3} = \frac{5}{6}$

15. $\frac{1}{6}$

17. (a) $\frac{3}{12}$
    (b) $\frac{3}{10}$
    (c) $\frac{3}{8}$

19. (a) $\frac{3}{4}$
    (b) $\frac{6}{4} = \frac{3}{2}$
    (c) $\frac{4}{12} = \frac{1}{3}$

Chapter 10 Number sentences

SECTION 1

1. It should balance. If it doesn't, check that the paper clips are the same weight. Slight variations will make a big difference.

2. (a) Balanced
   (b) Not balanced
   (c) Unknown: not enough information
   (d) Balanced
   (e) Unknown
   (f) Not balanced

5. (a) Balanced
   (b) Unknown
   (c) Not balanced

7. (a) $(1 \times 5) + (1 \times 10)$
   (b) $(3 \times 2) + (3 \times 5)$

8. (a) $(3 \times 6) = (1 \times 10) + (1 \times 8)$ or $(3 \times 6) = 10 + 8$

10. (a) True
    (b) False

11. (a) $10 + 6 = 2 \times \square$
    (b) $(4 \times \square) + 2 = 3 \times 10$

SECTION 2

1. (a) Yes   (b) Yes   (c) No
   (d) Yes   (e) No   (f) Yes

2. Answers will vary.

3. Answers will vary.

5. Truth sets for equations generally have one solution. Truth sets for inequalities usually involve a set of numbers. There are important exceptions to this rule which you will come across later on in the chapter.

7. (a) None   (b) Some
   (c) All   (d) Some

8. Answers will vary.

9. Answers will vary.

10. Answers will vary.

SECTION 3

1. $2 \times 3 = 6$

4. You can remove exactly the same number of paper clips from the same numbers on both sides of the balance without tipping the balance.

6. (a) $\square = 50$

   (b) $\triangle = 100$

   (c) $\square = 60$ and $\triangle = 2$ (Pairs of numbers can vary.)

   (d) $\varhexagon = $ any number

13. (a) $(8 \times \square) + 10 = 50; \quad 8 \times \square = 40$ (Subtract 10)

    (b) $(4 \times \square) + 8 = 32; \quad \square + 2 = 8$ (Divide by 4)

15. (a) $(4 \times \square) + 5 = 33; \quad 4 \times \square = 28; \quad \square = 7$

    (b) $(\square \times 6) + (\square \times 4) + 3 = 33; \quad (10 \times \square) \times 33; \quad 10 \times \square = 30; \quad \square = 3$

16. (a) $(5 \times \triangle) + 9 = 34; \quad 5 \times \triangle = 34 - 9$

    (b) $(6 \times \square) + 12 = 24; \quad 6 \times \square = 24 - 12$

17. (a) $7 \times \varhexagon = 21; \quad (7 \times \varhexagon) \div 7 = 21 \div 7; \quad \varhexagon = 3$

    (b) $4 \times \square = 28; \quad (4 \times \square) \div 4 = 28 \div 4; \quad \square = 7$

18. (a) $(3 \times \square) = 6; \quad (3 \times \square) \div 3 = 6 \div 3; \quad \square = 2$

    (b) $(5 \times \square) + 15 = 40; \quad 5 \times \square = 40 - 15; \quad (5 \times \square) \div 5 = 25 \div 5; \quad \square = 5.$

SECTION 4

1. (a) 27   (b) 1   (c) 16

   (d) 3   (e) 0   (f) 8

2. (a) $(24 \div 3) - (1 + 5) = 2$

   (b) $7 - 3 \times 4 + 5 = 2$ (Impossible!)

   (c) $15 - \left[3 \times (3 + 2)\right] = 0$

3. (a) 14

   (b) 14

Chapter 11 Equations and graphs

SECTION 1

3. C (8, 6); D (4, 10); E (9, 3); F (3, 7)

4. (a)

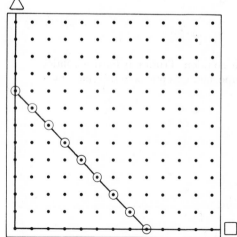

(b)

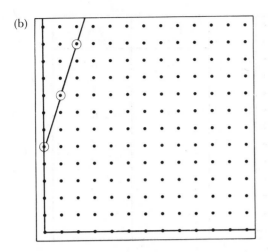

(c)

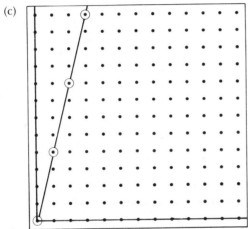

(d) The graph is a straight line.

5. (a) △

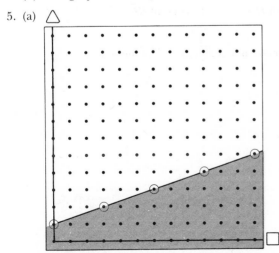

6. (a)

$\square$	$\triangle$
0	0
1	3
2	6
3	9
4	12

SECTION 2

2. (a) To the right one and up one
   (b) To the right three and up two
   (c) To the right one and down two
   (d) To the right one and up four

4. The number multiplied by the $\square$ is the slope.
$(a \times \square) + b = \triangle$ where a is the slope

6. Open sentences of the form $(a \times \square) + b = \triangle$ intersect the vertical axis at point b.

SECTION 3

1. Slope of $\overline{EF} = \frac{1}{2}$; slope of $\overline{GH} = \frac{^-3}{1}$; slope of $\overline{IJ} = \frac{^-3}{3} = ^-1$; slope of $\overline{KL} = \frac{2}{3}$

4. Slopes of parallel lines are equal.

7. The product of the slopes of two perpendicular line segments is $^-1$. One slope is the negative reciprocal of the other.

9. (a) $\frac{6}{6} = 1$
   (b) $\frac{5}{6}$
   (c) $\frac{4}{6} = \frac{2}{3}$

11. (a) $\frac{6}{6} = 1$
    (b) $\frac{6}{5}$
    (c) $\frac{6}{4} = \frac{3}{2}$

12. $\overline{OB}_7$ has no slope. A very large number.

14. $\square = 1$

SECTION 4

1. (2, 4)

3. (a)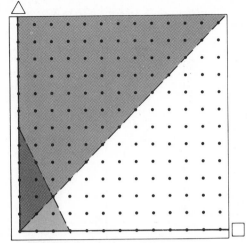

(b)

SECTION 5

2. (a)

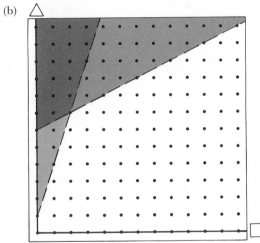

☐	△
0	8
1	3
2	0
3	¯1
4	0
5	3

(b)

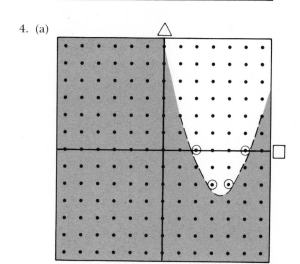

□	△
1	8
2	3
3	0
4	⁻1
5	0
6	3

4. (a)

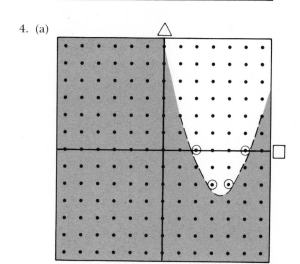

5. (a)

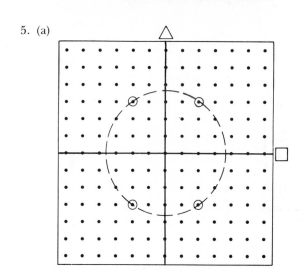

□	△
2	3
2	⁻3
⁻2	3
⁻2	⁻3
0	$\sqrt{13}$
0	⁻$\sqrt{13}$
$\sqrt{13}$	0
⁻$\sqrt{13}$	0

Chapter 12 The language of sets

SECTION 1

2. 8, 4, large and red, small and red, large and yellow, small and yellow, large and green, small and green, large and blue, small and blue

4. size and color; size and shape

R (square)	R (triangle)	R (circle)	R (parallelogram)
G (square)	G (triangle)	G (circle)	G (parallelogram)
Y (square)	Y (triangle)	Y (circle)	Y (parallelogram)
B (square)	B (triangle)	B (circle)	B (parallelogram)

5. The crossing: The figure at the crossing must be alike in 2 ways with 4 figures; yes; yes.

SECTION 2

2. Yes, no triangular figures are also square figures, and vice versa.
No, some green figures are also square figures and vice versa.

3. Yes, in the overlap. The figures in the set of green square figures belong to both of the other sets.

5. Answers will vary.

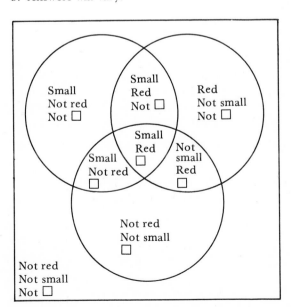

SECTION 3

2. {red figures}

4. The list method is good for sets with just a few elements; the rule method is good for sets with lots of elements.

5. (a) $\in$　　(b) $\subseteq$　　(c) $\subseteq$
   (d) $\subseteq$　　(e) $\subseteq$　　(i) $\in$
   (l) $\not\subseteq$

6. (a) {small, circular, figures}
   (b) {blue, triangular, figures}
   (c) {square, figures}
   (d) $\varnothing$

SECTION 4

2. Yes; yes; every combination of elements in the table generates one of the elements in the set.

4. Yes; $(R \cup X) \cup Y = Y = R \cup (X \cup Y)$
   Yes; $X \cup Y = Y = Y \cup X$

6. Only for $\cup$; only for $\varnothing$

8. Yes; $R \cap (X \cup C) = X = (R \cap X) \cup (R \cap C)$

SECTION 5

1. (a) 16　　(b) 8　　(c) 8
   (d) 4　　(e) 4　　(f) 4
   (g) 2

3. Whenever the sets are disjoint, i.e., whenever the intersection of the sets is $\varnothing$.

SECTION 6

1. (a) {blue figures} $\cap$ {triangular figures} = {blue triangular figures}

2. (a) A or not A is all 32 attribute shapes.
   (b) A and not A is nothing.

Chapter 13　Probability and statistics

SECTION 1 Answers will vary.

SECTION 2

Experiment 1: Answers will vary.

Experiment 2: Answers will vary in Table 3.
   (a) Yes
   (b) It is conjectured that some digits are more likely to occur than others as the first digit of a telephone number.

SECTION 3

1. $\frac{1}{4}, \frac{1}{4}$

3. (a) $\frac{1}{2}$
   (b) $\frac{3}{10}$
   (c) $\frac{1}{5}$

5. (a) $\frac{1}{10}$
   (b) $\frac{7}{10}$
   (c) $\frac{3}{10}$
   (d) $\frac{5}{10}$

SECTION 4

Experiment 3: Answers will vary.

Experiment 4: Answers will vary.
   (c) The sum of the probabilities of all possible events of any experiment is 1.

Experiment 5: Answers will vary.
   (c) Most people will get tails 60 to 90 percent of the time, depending on the age of the penny.

Experiment 7: 16 or 17
   (c) The theoretical probability of rolling a 3 with a single die is $\frac{1}{16}$.
   (d) Theoretical probabilities are based on a "perfect" model of a situation. 100 trials is not necessarily enough trials to accurately determine the theoretical probability.

Chapter 14 Metric measure

SECTION 1

3. They were both "correct" although their "spans" were of different sizes.

4. If I asked a friend to get me 50 cubits of rope, I either don't *care* about the exact length or else I don't know much about measurement. My friend's cubit may be longer or shorter than mine.

SECTION 2

2. Probably not. Each person's body is different in size.

6. Estimates will vary. Measurements are 4.8 cm, 3.5 cm, 17.9 cm, 12.1 cm.

SECTION 3

1. 8 cm $\times$ 8.5 cm; $A = 64$ cm^2
   13 cm $\times$ 5 cm; $A = 65$ cm^2

6. Find out how many "hands" it takes to cover your entire body.

SECTION 4

1. (a) $\frac{1}{10}$ m $\times$ $\frac{1}{10}$ m $\times$ $\frac{1}{10}$ m   (b) 10 cm $\times$ 10 cm $\times$ 10 cm
   (c) $\frac{1}{1000}$ m^3 = 0.001 m^3 (d) 1000 cm^3

3. Answers will vary: *Example*: weight = 220 lb
                      $220 \div 2.2 = 100$ kg
                      1 kg water = 1 liter
Hence your volume is about 100 liters if you weigh 220 lb.

SECTION 5 In this section answers will vary. The important concept is your procedures for estimating and measuring and the precision that you used.

SECTION 6

2. The similarities of the measures for length, volume, and mass and the ease with which we can convert among these measures is obviously a human invention.

4. Decimal arithmetic is based on multiples of 10, just as the measures of the metric system.

6. kilo   = 1000
   hecto = 100
   deka   =   10
   milli   = $\frac{1}{100}$
   centi   = $\frac{1}{100}$
   deci   = $\frac{1}{10}$

Chapter 15 Geometry

SECTION 1

2. Conjecture: $m(\overline{AX}) + m(\overline{CX}) = m(\overline{BX})$

4. Conjecture: They are proportional.

6. Conjecture: $m(\overline{AP}) \times m(\overline{PB}) = m(\overline{CP}) \times m(\overline{PD})$

8. Conjecture: $m(\overline{AB}) \times m(\overline{AC}) = m(\overline{AD}) \times m(\overline{AE})$

10. Conjecture: $\dfrac{m(\overline{AD})}{m(\overline{DB})} \dfrac{m(\overline{BE})}{m(\overline{EC})} \dfrac{m(\overline{CE})}{m(\overline{FA})} = 1$

12. Conjecture: The lines are concurrent, that is, they intersect in a single point.

14. Conjecture: The measure of any inscribed angle which intercepts a semicircle is 90°. They are right angles.

SECTION 2

2.

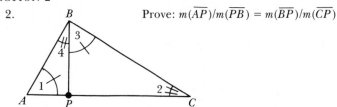

Prove: $m(\overline{AP})/m(\overline{PB}) = m(\overline{BP})/m(\overline{CP})$

Statement	Reason
(1) $m\angle 1 + m\angle 4 = 90°$ $m\angle 1 + m\angle 2 = 90°$ $m\angle 2 + m\angle 3 = 90°$	(1) Sum of m ∠'s of a △ = 180° given that each △ is a right △.
(2) $m\angle 2 = m\angle 4$ $m\angle 1 = m\angle 3$	(2) From step 1, equations 1 and 2. From step 1, equations 2 and 3.
(3) $\triangle ABP \sim \triangle BCP$	(3) Similar triangles.
(4) $\dfrac{m(\overline{AP})}{m(\overline{PB})} = \dfrac{m(\overline{BP})}{m(\overline{CP})}$	(4) Corresponding sides of similar triangles are proportional.

3.

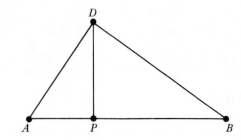

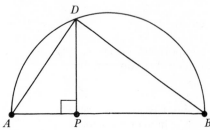

Prove: $m(\overline{AP}) \times m(\overline{PB}) = m(\overline{PD})^2$

Statement	Reason
(1) $\angle ADB$ is a right angle	(1) Given
(2) $\dfrac{m(\overline{AP})}{m(\overline{PD})} = \dfrac{m(\overline{PD})}{m(\overline{PB})}$	(2) Proof for 2 problem
(3) $m(\overline{AP}) \times m(\overline{PB}) = m(\overline{PD})^2$	(3) Property of proportions

Index

Signed number, 166-167
  negative number, 166
  positive number, 166
Similar, 52
Slope, 258, 266
Square, 30, 33
Square root, 45
Square unit, 30
Standard chain, 94, 130
Structural model, 29
Subset, 297
Subtend, 361
Subtraction, 109
Sum, 107
Symmetrical, 20

Tangent, 355
Theoretical probability, 310
Transformation, shear, 39
Transversal, 356
Triangle, 29, 37
  acute, 30

altitude, 352
base of, 39
equilateral, 30, 349
height, 39
hypotenuse, 43
Isosceles, 30, 39
legs of, 43
Pascal's, 21
Pythagorean theorem, 43
right, 30, 42, 342
scalene, 30
Truth set, 232

Union, of sets, 298
Uniqueness, 177
Unit of measure, 30
Universal set, 292

Variable, 71
Vertex, 17
Vertical axis, 277